Akira Namatame, Satoshi Kurihara and Hideyuki Nakashima (Eds.)

Emergent Intelligence of Networked Agents

# Studies in Computational Intelligence, Volume 56

Editor-in-chief
Prof. Janusz Kacprzyk
Systems Research Institute
Polish Academy of Sciences
ul. Newelska 6
01-447 Warsaw
Poland
*E-mail:* kacprzyk@ibspan.waw.pl

Akira Namatame
Satoshi Kurihara
Hideyuki Nakashima
(Eds.)

# Emergent Intelligence of Networked Agents

With 105 Figures and 28 Tables

 Springer

Akira Namatame
Department of Computer Science
National Defense Academy
1-10-20, Hashirimizu
Yokosuka, 239-8686
Japan
E-mail: nama@nda.ac.jp

Hideyuki Nakashima
Future University-Hakodate
General co-chair of AAMAS-06
116-2 Kamedanakano-cho Hakodate
Hokkaido 041-8655
Japan
E-mail: h.nakashima@fun.ac.jp

Satoshi Kurihara
Graduate School of
Information Science and Technology
Osaka University
8-1 Mihogaoka
Ibaraki, Osaka, 567-0047
Japan
E-mail: kurihara@ist.osaka-u.ac.jp

Library of Congress Control Number: 2007921689

ISSN print edition: 1860-949X
ISSN electronic edition: 1860-9503
ISBN-10   3-540-71073-6 Springer Berlin Heidelberg New York
ISBN-13   978-3-540-71073-8 Springer Berlin Heidelberg New York

Springer is a part of Springer Science+Business Media
springer.com
© Springer-Verlag Berlin Heidelberg 2007

Cover design: deblik, Berlin
Typesetting by the editors using a Springer LATEX macro package
Printed on acid-free paper      SPIN: 11679851      89/SPi      5 4 3 2 1 0

# Preface

Recently, the study of intelligence emerged from interactions among many agents has been popular. In this study it is recognized that a network structure of the agents plays an important role. The current state-of-the art in agent-based modeling tends to be a mass of agents that have a series of states that they can express as a result of the network structure in which they are embedded. Agent interactions of all kinds are usually structured with complex networks. Research on complex networks focuses on scale-freeness of various kind of networks.

Computational modeling of dynamic agent interactions on richly structured networks is important for understanding the sometimes counter-intuitive dynamics of such loosely coupled systems of interactions. Yet our tools to model, understand, and predict dynamic agent interactions and their behavior on complex networks have lagged far behind. Even recent progress in network modeling has not yet offered us any capability to model dynamic processes among agents who interact at all scales on such as small-world and scale-free networks. Generally the high-dimensional, non-linear nature of the resulting network-centric multi-agent systems makes them difficult or impossible to analyze using traditional methods. Agents follow local rules under complex network constraints. The idea of combining multi-agent systems and complex networks is also particularly rich and fresh to foster the research on the study of very large-scale multi-agent systems.

We intend to turn this into an engineering methodology to design complex agent networks. Multi-agent network dynamics involves the study of many agents, constituent components generally active ones with a simple structures and whose behavior is assumed to follow local rules, and their interactions on complex network. A basic methodology is to specify how the agents interact, and then observe emergent intelligence that occur at the collective level in order to discover basic principles and key mechanisms for understanding and shaping the resulting intelligent behavior on network dynamics.

The volume contains refereed papers addressing various important topics that aims at the investigation of emergent intelligence on networked agents.

Especially most papers highlight on the topics such "network formation among agents", "influence of network structures on agents", "network-based collective phenomena and emergent intelligence on networked agents".

The selected papers of this volume were presented at the Workshop on Emergent Intelligence of Networked Agents (WEIN 06) at the Fifth International Joint Conference on Autonomous Agents and Multi-agent Systems (AAMAS 2006), which was held at Future University, Hakodate, Japan, from May 8 to 12, 2006. WEIN 06 is concerned with emergence of intelligent behaviors over networked agents and fostering the formation of an active multi-disciplinary community on multi-agent systems and complex networks. We especially intended to increase the awareness of researchers in these two fields sharing the common view on combining agent-based modeling and complex networks in order to develop insight and foster predictive methodologies in studying emergent intelligence on of networked agents. From the broad spectrum of activities, leading experts presented important paper and numerous practical problems appear throughout this book. We invited high quality contributions on a wide variety of topics relevant to the wide research areas of multi-agent network dynamics. We especially covered in-depth of important areas including: Adaptation and evolution in complex networks, Economic agents and complex networks, Emergence in complex networks, Emergent intelligence in multi-agent systems, Collective intelligence, Learning and evolution in multi-agent systems, Web dynamics as complex networks, Multi-agent based supply networks, Network-centric agent systems, Scalability in multi-agent systems, Scale-free networks, Small-world networks.

We could solicit many high quality papers that reflect the result of the growing recognition of the importance of the areas. All papers have received a careful and supportive review, and we selected 19 papers out of 31 papers. The contributions were submitted as a full paper and reviewed by senior researchers from the program committee. All authors revised their earlier versions presented at the workshop with reflecting criticisms and comments received at the workshop. The editors would like to thank the program committee for the careful review of the papers and the sponsors and volunteers for their valuable contribution. We hope that as a result of reading the book you will share with us the intellectual excitement and interest in this emerging discipline. We also thank the many other referees who generously contributed time to ensure the quality of the finished product.

## Workshop Organizers

Satoshi Kurihara, Osaka University, Japan
Hideyuki Nakashima, Future University Hakodate, Japan
Akira Namatame, National Defense Academy, Japan, Workshop Chair

# International Steering Committee

Robert Axtell, Brookings Institution, and Santa Fe Institute, USA
Giorgio Fagiolo, University of Verona, Italy
Satoshi Kurihara, Osaka University, Japan
Hideyuki Nakashima, Future University Hakodate, Japan
Akira Namatame, National Defense Academy, Japan

# Scientific Program Committee

Robert Axtell, Santa Fe Institute, USA
Sung-Bae Cho, Yosei University, Korea
Giorgio Fagiolo, University of Verona, Verona, Italy
Kensuke Fukuda, National Institute of Informatics (NII), Japan
David Green, Monash University, Australia
Yukio Hayashi, Japan Advanced Institute of Science and Technology
    (JAIST), Japan
Dirk Helbing, Dresden Technical University, Germany
Kiyoshi Izumi, National Institute of Advanced Industrial Science and Tech-
    nology (AIST), Japan
Taisei Kaizoji, International Christian University (ICU), Japan
Hidenori Kawamura, Hokkaido University, Japan
Satoshi Kurihara, Osaka University, Japan
Yutaka Matsuo, National Institute of Advanced Industrial Science and
    Technology (AIST), Japan
Peter Mika, Free University of Amsterdam, Netherlands
Hideyuki Nakashima, Future University - Hakodate, Japan
Akira Namatame, National Defense Academy, Japan
Denis Phan, University of Rennes, France
Jon Sakker, Australian Defense Academy, Australia
Frank Schweitzer, ZTH, Switzerland
Wataru Souma,Advanced Telecommunications Research Institute Interna-
    tional (ATR), Japan
David Wolpert, NASA Ames Research Center, USA

Tokyo,                                          *Akira Namatame*
November 2006                              *Satoshi Kurihara*
                                              *Hideyuki Nakashima*

# Contents

# Incremental Development of Networked Intelligence in Flocking Behavior

Masaru Aoyagi and Akira Namatame

Dept. of Computer Science, National Defense Academy
{g45074, nama}@nda.ac.jp

**Abstract.** In this paper, we present the model for incremental development success of emerging intelligence over networked agents which from purposive flocking behavior. We especially discuss how to give the agents the ability to follow spatial restriction while keeping the complexity low enough to still allow for real-time moving creatures. Our method involves by extending Reynolds' flocking algorithm. We aim to guide a collection of moving agents to fly in natural-looking paths, and travel smoothly in 3D space, with speed regulation in curve.

## 1 Introduction

To creatures a large number of objects by hand would be a tedious job. To make the job easier, we would like to try to automate as much of the process as possible. For the case of flocks or herds of creatures, Craig Reynolds [1, 2] introduced a simple agent-based approach to animate a flock of creatures through space. In this method, each creature makes its own decisions on how to move, according to a small number of simple rules that consider the neighboring members of the flock. Reynolds suggested that further modifications could lead to a herd model by giving the creatures the ability to follow spatial restriction.

The challenge is how to give the creatures the ability to follow spatial restriction while keeping the complexity low enough to still allow for real-time simulation of the herd. We would like the herd to move in natural-looking paths. Also, we would like the creatures to travel smoothly in 3D space, with speed regulation in curve.

Increasing use of autonomous unmanned air vehicles in a variety of civil and military applications is putting increasing pressure on traditional airspace management capabilities. One solution to this problem is decentralize the management function by delegating control such that individual aircraft manage their own airspace by negotiating with neighboring aircraft, i.e. free-flight. For example, two aircraft on a potential collision course will negotiate changes to their respective flight plans to remove the risk of collision. This negotiation takes place

M. Aoyagi and A. Namatame: *Incremental Development of Networked Intelligence in Flocking Behavior*, Studies in Computational Intelligence (SCI) **56**, 1–12 (2007)
www.springerlink.com

independently of a ground air traffic controller. However, for very high airspace densities, real time negotiation of alternative flight paths becomes impractical and it becomes necessary to impose a set of flight rules that minimizes the probability of collisions a priori.

In nature, aggregations of large numbers of mobile organisms are also faced with the problem of organizing themselves efficiently. This selective pressure has led to the evolution of behavior such as flocking of birds, swarming of insects, herding of land animals and schooling of fish. The reasons why organisms form flocks are varied and include protection from predation, improved food search and improved social cohesion. However, the actual dynamics of the flocking behavior are essentially constrained by the dynamics of the individual organisms and the flock is relatively limited in the types of behavior it can exhibit. This gives a flock of a given organism, be it fish or bird, its characteristic look and feel.

## 2 Literatures on Flocking Behavior

A flock may be loosely defined as a clustered group of individuals with a common velocity vector. Note that flocking of aircraft is different from formation flying. In the latter, aircraft are arranged according to predefined relationships that generally remain fixed during the flight. With flocking flight, there are no predefined relation-ships and the flock members may constantly change their position within the group. The fixed relationships within aircraft formations make them relatively difficult to maneuver, whereas the fluid nature of a flock allows relatively rapid changes in flock direction.

The aim of the present work is to demonstrate flocking behavior of a group of simulated unmanned air vehicles. Of particular interest are the development of meaningful statistical metrics that usefully quantify flocking behavior and the investigation of the relationship between rule weighting and flocking behavior.

Underlying this type of motion control (steering objects around obstacles toward goals) is the concept of self-directed action. Rather than the traditional view of inert geometrical objects being moved according to a centralized, pre-existing plan (the script specified by the animator), here we consider the objects to be active independent entities capable of directing their own motion. They are not merely inert geometrical models but rather "self-motivated" behavioral models, which also happen to have a geometric component. Actually, most convenient are hybrid objects that can be directed in general terms from an animator's script but that can handle details of their own behavior as required. Specifically, the animator can set up goals for the behavioral objects, but the objects themselves will work out the specifics, such as the exact path that they take to avoid collisions with obstacles. Furthermore, the "guiding hand" of the animator is not required; behavioral objects are perfectly happy to merely wander around on their own, ad libbing their roles as "extras."

Formation behaviors in nature, like flocking and schooling, benefit the animals that use them in various ways. Each animal in a herd, for instance, benefits by minimizing its encounters with predators [3]. By grouping, animals also combine their sensors to maximize the chance of detecting predators or to more efficiently forage for food. Studies of flocking and schooling show that these behaviors emerge as a combination of a desire to stay in the group and yet simultaneously keep a separation distance from other members of the group [4]. Since groups of artificial agents could similarly benefit from formation tactics, robotics researchers and those in the artificial life community have drawn from these biological studies to develop formation behaviors for both simulated agents and robots. Approaches to formation generation in robots may be distinguished by their sensing requirements, their method of behavioral integration, and their commitment to preplanning. A brief review of a few of these efforts follows.

The viability of obtaining coherent flocking behavior from simple rules was first demonstrated by Reynolds [1]. The primary application for this work was in developing realistic motions of groups of 'actors' in computer animation. Conventionally, each 'actor' (known generically as a 'boid' in flocking work) has a scripted path predetermined by the animator. For large numbers of boids, for example a flock of birds, the process was cumbersome and did not produce realistic results. This led to the use of relatively simple flocking rules that would automatically govern the dynamic behavior.

Improvements to this approach have recently been made by Tu and Terzopoulos and separately by Brogan and Hodgins. Tu and Terzopoulos [5] developed more realistic simulated fish schooling by accurately modeling the animals' muscle and behavioral systems. Brogan and Hodgins [6] developed a system for realistically animating herds of one-legged agents using dynamical models of robot motion. Both results are more visually realistic than Reynolds' because they simulate the mechanics of motion; Reynolds' approach utilized particle models only.

The individual components of Reynolds' flocking and Brogan's herding behaviors are similar in philosophy to the motor schema paradigm used here, but their approaches are concerned with the generation of visually realistic flocks and herds for large numbers of simulated animals, a different problem domain than the one this article addresses. In contrast, our research studies behaviors for a small group (up to four) of mobile robots, striving to maintain a specific geometric formation.

The dynamics and stability of multi-robot formations have drawn recent attention [7], [8]. Wang [7] developed a strategy for robot formations where individual robots are given specific positions to maintain relative to a leader or neighbor. Sensory requirements for these robots are reduced since they only need to know about a few other robots. Wang's analysis centered on feedback control for formation maintenance and stability of the resulting system. It did not include integrative strategies for obstacle avoidance and navigation. In work by Chen and Luh [8] formation generation by distributed control is demonstrated. Large groups of robots are shown to cooperatively move in various geometric formations. Chen's research also centered on the analysis of group dynamics and stability, and does not provide for obstacle avoidance. In the approach forwarded in this article, geometric formations are specified in a similar manner, but formation behaviors are fully integrated with obstacle avoidance and other navigation behaviors.

Mataric has also investigated emergent group behavior [9], [10]. Her work shows that simple behaviors like avoidance, aggregation and dispersion can be combined to create an emergent flocking behavior in groups of wheeled robots. Her research is in the vein of Reynolds' work in that a specific agent's geometric position is not designated. The behaviors described in this article differ in that positions for each individual robot relative to the group are specified and maintained.

Related papers on formation control for robot teams include [11], [12], [13], [14]. Parker's thesis [11] concerns the coordination of multiple heterogeneous robots. Of particular interest is her work in implementing "bounding over watch," a military movement technique for teams of agents; one group moves (bounds) a short distance, while the other group over watches for danger. Yoshida [12], and separately, Yamaguchi [13], investigate how robots can use only local communication to generate a global grouping behavior. Similarly, Gage [14] examines how robots can use local sensing to achieve group objectives like coverage and formation maintenance. Parker simulates robots in a line-abreast formation navigating past waypoints to a final destination [15]. The agents are programmed using the layered subsumption architecture [5]. Parker evaluates the benefits of varying degrees of global knowledge in terms of cumulative position error and time to complete the task. Tucker and Ronald [17] presented reactive behavior for formation keeping and demonstrated successfully outdoors on cars.

## 3 A Boids Model Simulation for Flocking Behavior

A method is presented for flocking behavior of creatures, birds, fishes and so on, that can form herds by evading obstacles in airspace, terrain or ocean floor topography in 3D space while being efficient enough to run in real-time. This method involves making modifications to Reynolds' flocking algorithm [2] as following.

- Cohesion: steer to move toward the average position of local flockmates
- Separation: steer to avoid crowding local flockmates
- Alignment: steer towards the average heading of local flockmates

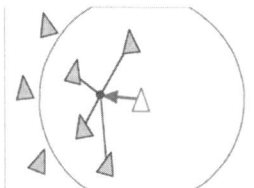   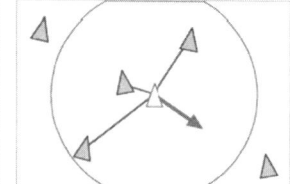   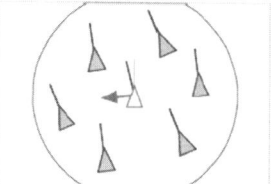

**Fig. 1.** Cohesion          **Fig. 2.** Separation          **Fig. 3.** Alignment

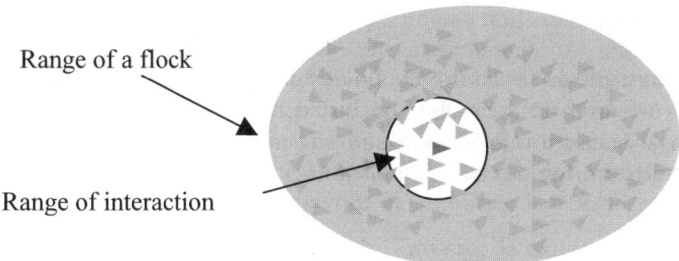

Range of a flock

Range of interaction

**Fig. 4.** Massively Flocking

The agent is individual and it is behavior determines how an agent reacts to other characters in its local neighborhood, as shown grey area in Fig. 1, 2 and 3. Agents outside of the local neighborhood are ignored. A Flock often consists of multi local interactions of each agent. We define that the flock consists of multi local interactions as a massively flocking. (Fig. 4)

Cohesion behavior gives an agent (outlined triangle located in the centre of the diagram) the ability to cohere with (approach and form a group with) other nearby agents. See Fig. 1. Steering for cohesion can be computed by finding all characters in the local neighborhood (as described above for separation), computing the average position (or center of gravity) of the nearby characters. The steering force can applied in the direction of that average position (subtracting our character position from the average position, as in the original agents model). The flocking rules used in the present work are illustrated schematically. The cohesion rule, acts such that the active flock member (outlined triangle located in the centre of the diagram) tries to orient its velocity vector in the direction of the centroid (average spatial position) of the local flock. The degree of locality of the rule is determined by the sensor range of the active flock member, represented diagrammatically by the light colored circle. Note that for effective cohesion it is necessary to vary the speed of the active flock member as well as it's heading such that the speed of agents far from the flock centroid is increased. This is referred to as speed cohesion and allows wayward agents to catch up with the rest of the flock.

Separation behavior gives an agent the ability to maintain a certain separation distance from others nearby. This can be used to prevent agents from crowding together. To compute steering for separation, first a search is made to find other characters within the specified neighborhood. This might be an exhaustive search of all characters in the simulated world, or might use some sort of spatial partitioning or caching scheme to limit the search to local characters. For each nearby character, a repulsive force is computed by subtracting the positions of our character and the nearby character, normalizing, and then applying a $1/d$ weighting. d is defined as distance between our character and the nearby character. Note that $1/d$ is just a setting that has worked well, not a fundamental value. These repulsive forces for each nearby character are summed together to produce the overall steering force as show in Fig. 2.

Alignment behavior gives a character the ability to align itself with (that is, head in the same direction and/or speed as) other nearby characters, as shown in Fig. 3. Steering for alignment can be computed by finding all characters in the local neighborhood (as described above for separation), averaging together the velocity (or alternately, the unit forward vector) of the nearby characters. This average is the desired velocity, and so the steering vector is the difference between the average and our agent's current velocity (or alternately, its unit forward vector). This behavior will tend to turn our character so it is aligned with its neighbors.

Simulations show that reasonable flocking behavior can be obtained using just cohesion and alignment rules. Left unchecked, the cohesion rules will tend to lead to flock overcrowding. To balance this, a separation rule is used, where the active flock member tries to translate away from the local flock centroid. Note that for effective flocking behavior, the sensor range of the cohesion rules will generally be much larger than the separation rule, i.e. cohesion acts at a global level whereas separation works locally.

Each agent has direct access to the whole scene's geometric description, but flocking requires that it react only to flockmates within a certain small neighborhood around itself. The neighborhoods are characterized by a distance (measured from the center of the agent) and an angle, measured from the agent's direction of flight. Flockmates outside this local neighborhood are ignored. The neighborhood could be considered a model of limited perception but it is probably more correct to think of it as defining the region in which flockmates influence an agents steering. The modifications use only local properties of the spatial restriction, and thus have low complexity and wide application. The flocking algorithm with these modifications produces naturally behaving herds that follow the spatial restriction. The evading obstacles rule added to the flocking algorithm has a constant parameter that can be adjusted to produce different behaviors.

The flocking algorithm works as follows: For a given agent, centroids are calculated using the sensor characteristics associated with each flocking rule. Next, the velocity vector the given agent should follow to enact the rule is calculated for each of the rules. These velocity vectors are then weighted according to the rule strength and summed to give an overall velocity vector demand. Finally, this velocity vector demand is resolved in to a heading angle, pitch attitude and speed demand, which is passed to the control system. The control system then outputs an actuator vector that alters the motion of the aircraft in the appropriate manner.

## 4 The Steady-State Analysis of Flocking Behavior

Each agent recognizes two physical values (Fig. 5). One is the position to flockmates from the agent; the other is the relative velocity of flockmates. Agent i acquires the flockmate agent j in visual sensor range. Agent i can recognize vector $\vec{d}_{ij}$ that is distance to the flockmate agent j. It also recognizes vector

$\vec{v}_{ij} = d\vec{d}_{ij}/dt$ that is relative velocity. Cohesion force vector $\vec{F}_{ci}$, separation force vector $\vec{F}_{si}$ and alignment force vector $\vec{F}_{ai}$ are defined form these two physical values, such that

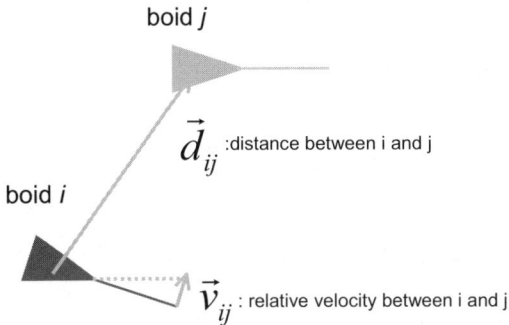

**Fig. 5.** Relative position and velocity vector that a agent recognizes

$$\vec{F}_{ci} = w_{ci}\frac{\sum\limits_{j}^{n_i}\vec{d}_{ij}}{\left|\sum\limits_{j}^{n_i}\vec{d}_{ij}\right|}, \quad \vec{F}_{si} = -w_{si}\frac{\sum\limits_{j}^{n_i}\vec{d}_{ij}}{\left|\sum\limits_{j}^{n_i}\vec{d}_{ij}\right|^2}, \quad \vec{F}_{ai} = w_{ai}\frac{\sum\limits_{j}^{n_i}\vec{v}_{ij}}{\left|\sum\limits_{j}^{n_i}\vec{v}_{ij}\right|}, \tag{1}$$

where coefficient $w_{ci}$, $w_{si}$ and $w_{ai}$ is a weight each force and positive. Then, flocking force vector $\vec{F}_{fi}$ expression is defined as liner combination of cohesion, separation and alignment force.

$$\vec{F}_{fi} = \vec{F}_{ci} + \vec{F}_{si} + \vec{F}_{ai} = \left(w_{ci} - \frac{w_{si}}{\left|\sum\limits_{j}^{n_i}\vec{d}_{ij}\right|}\right)\frac{\sum\limits_{j}^{n_i}\vec{d}_{ij}}{\left|\sum\limits_{j}^{n_i}\vec{d}_{ij}\right|} + w_{ai}\frac{\sum\limits_{j}^{n_i}\vec{v}_{ij}}{\left|\sum\limits_{j}^{n_i}\vec{v}_{ij}\right|} \tag{2}$$

If

$$\vec{F}_{fi} = \vec{0}, \tag{3}$$

then, the flock moving becomes steady-state. In this case, both first and second term in Eq. (2) has to equals zero.

The condition of first term in Eq. (2) is

$$\left| \sum_{j}^{n_i} \vec{d}_{ij} \right| = \frac{w_{si}}{w_{ci}} . \tag{4}$$

Eq. (4) shows the distance between agent and flockmates in steady-state. If $w_{si}$ is smaller or $w_{ci}$ is larger, then the distance $\vec{d}_{ij}$ becomes shorter. The steady-state condition of second term, alignment term, in Eq. (2) is

$$\sum_{j}^{n_i} \vec{v}_{ij} = \sum_{j}^{n_i} \frac{d\vec{d}_{ij}}{dt} = \vec{0} . \tag{5}$$

Eq. (5) shows the velocity of both agent and flockmates is same in steady-state. If $w_{ai}$ is larger, then recognizes vector $\vec{v}_{ij} = d\vec{d}_{ij}/dt$ comes to zero more quickly, and the velocity of agent comes to the velocity of flockmates more quickly.

As each agent heads for a destination, they have two parallel intentions. The intentions are both flocking behavior and heading for destination. We combine the two intentions with a "Probabilistic Method". The method is that agent remember to head for destination by probability p. Total force of the whole flock $\vec{F}_{total}$ is

$$\vec{F}_{total} = \vec{F}_f + p\vec{F}_d = \frac{1}{n} \left\{ \sum_{i}^{n} \vec{F}_{fi} + p \sum_{i}^{n} \vec{F}_{di} \right\} . \tag{6}$$

where $\vec{F}_f$ is average flocking behavior force of agents, $\vec{F}_{di}$ is the force of intention that a agent i wants to head for destination. $\vec{F}_d$ is average force that agents want to head for destination.

## 5 Environmental setting for Intelligence Development of Flocking Behavior

3D simulation has to be for flying behavior. We develop 3D simulation using Java 3D. Java 3D is an addition to Java for displaying three-dimensional graphics.

Sometimes, unmanned vehicles fly through spatial restriction. There will be the unmanned vehicle's missions of flying low in the mountain area. We study that unmanned aerial vehicles fly through spatial restriction as they continue flocking. Then, we challenged 3 problems as follows.

**Environmental Complexity Level 1:**    "How to keep a flock by blocking an obstacle"

**Environmental Complexity Level 2:**    "Break away from dead lock and advanced about a destination"

A flock often breaks out by avoiding an obstacle (Fig. 6). We improve composition of flocking rules, cohesion, separation and alignment as an agent face to an obstacle. That method is called as "Re-combination 3 Local Behavioral Rules" in this paper. We also study about flocking behavior as agent has parallel intentions. Fig. 7 is this case. In this case, agent has two intentions. They are flocking behavior and heading for destination. We combine the two intentions with a probabilistic method. Table 1 shows conditions of simulation.

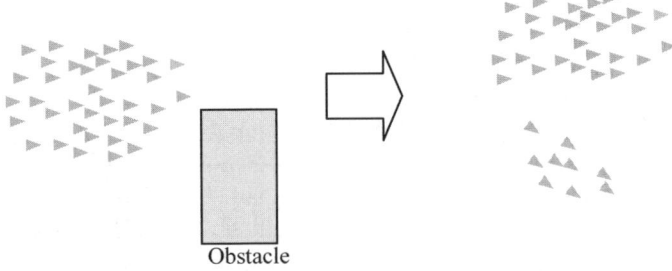

**Fig. 6.** A flock breaks up by avoiding an obstacle

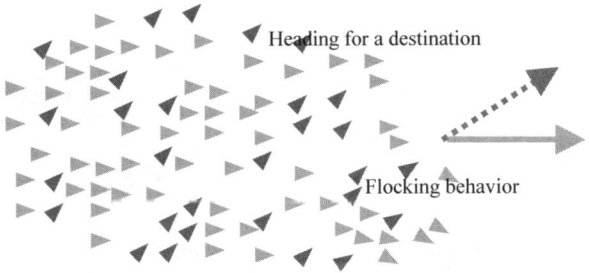

**Fig. 7.** Parallel intentions

## 6 Simulation Results

We show is the simulation result of Environmental Level 1 in Fig. 8. The situation is set as follows. There is a big obstacle in front of agents. A flock is closing to it. Each agent in the flock wants to avoid an obstacle. In the result without re-combination 3 local behavior rules, a red circle shows 4 flock members that have disjoined. The result without re-combination 3 local behavior rules shows flock members don't disjoined.

The simulation result of Environmental Level 2 is shown in Fig. 9. Red sphere is a destination. As agents always have the intention that go toward a destination and a box type obstacle stands in the way. Then the agents deadlock at the edge of box. As with probabilistic method, the flock breaks away form a box obstacle and goes toward the destination.

**Table 1.** Parameters in Simulation

| Parameter | Definition | Value | |
|---|---|---|---|
| N | The number of agents | 100 | |
| p | The probability of remembrance | 0.15 | |
| $R_v$ | The range of visual sensor for neighborhood | 3.0 | |
| $w_c$ | Cohesion weight | usually | 0.20 |
| | | in the act of avoiding | 0.32 |
| $w_s$ | Separation weight | usually | 0.20 |
| | | in the act of avoiding | 0.08 |
| $w_a$ | Alignment weight | usually | 0.20 |
| | | in the act of avoiding | 0.32 |

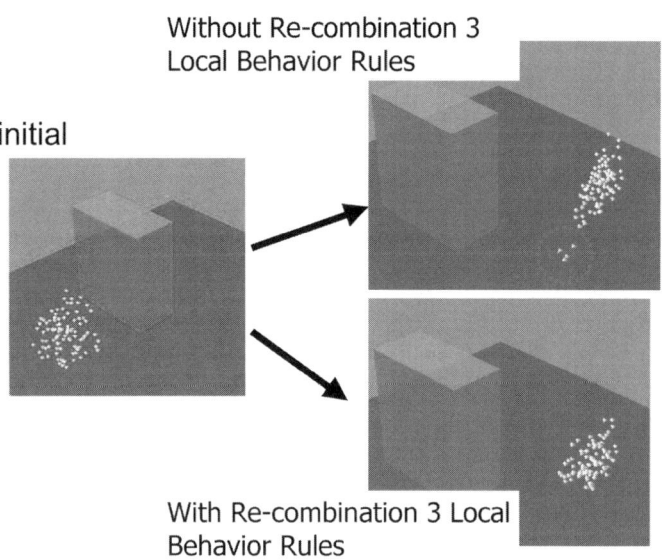

**Fig. 8.** Simulation results of avoiding an obstacle

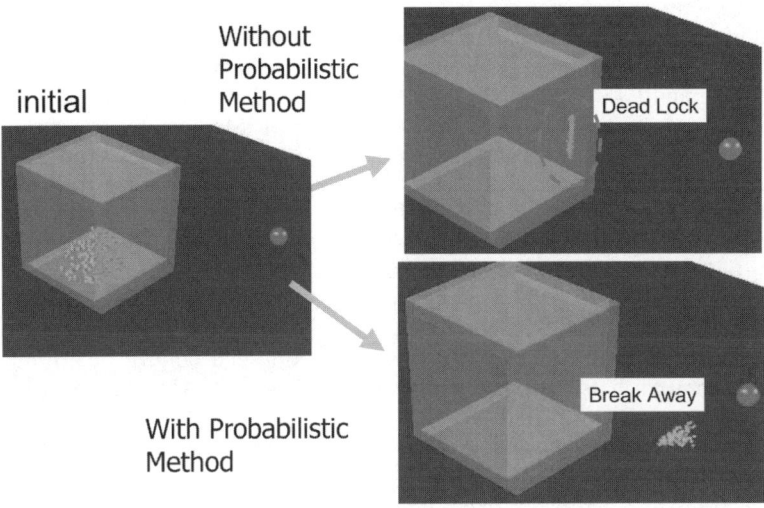

**Fig. 9.** Simulation results of flying through a box-like obstacle and toward a destination

## 7 Conclusion and Future Works

Flocking offers a potentially simple and efficient way of managing the flight paths of a large number of small autonomous UAVs such that the risk of collision and/or the need for evasive maneuvers is reduced. The way in which flocking rules are implemented depends strongly on the nature of the flight control system available on the target flight vehicle. We have two type Environmental Levels. First we challenged the problem how the flock don't disjoin when avoiding an obstacle. We have suggested "Re-combination 3 Local Behavior Rules" as a solution for this problem. Next we have challenged the problem how an agent have combined Intention both flocking behavior and go toward destination. We have suggested "Probabilistic Method" as a solution for this problem.

## References

1  Reynolds, C. W. (1987) "Flocks, Herds, and Schools: A Distributed Behavioral Model, in Computer Graphics," 21(4) (SIGGRAPH'87 Conference Proceedings), pp. 25/34
2  Reynolds, C. W. (1999) "Steering Behaviors For Autonomous Characters," in the proceedings of Game Developers Conference 1999 held in San Jose,

California. Miller Freeman Game Group, San Francisco, California. pp. 763/782

3   Veherencamp, S. L. (1987) "Individual, kin, and group selection," in Handbook of Behavioral Neurobiology, Volume 3: Social Behavior and Communication, P. Marler and J. G. Vandenbergh, Eds. New York: Plenum, pp. 354/382

4   Cullen, J. M., E. Shaw, and H. A. Baldwin (1965) "Methods for measuring the three-dimensional structure of fish schools," Animal Beh., vol. 13, pp. 534/543

5   Tu, X. and D. Terzopoulos (1994) "Artificial fishes: Physics, locomotion, perception, behavior," in Proc. SIGGRAPH 94 Conf., Orlando, FL, pp. 43/50

6   Brogan, D. C. and J. K. Hodgins (1997) "Group behaviors for systems with significant dynamics," Auton. Robots, vol. 4, no. 1, pp. 137/153

7   Wang, P. K. C. (1991) "Navigation strategies for multiple autonomous robots moving in formation," J. Robot. Syst., vol. 8, no. 2, pp. 177/195

8   Chen, Q. and J. Y. S. Luh (1994) "Coordination and control of a group of small mobile robots," in Proc. IEEE Int. Conf. Robot. Automat., San Diego, CA, 1994, pp. 2315/2320

9   Mataric, M. (1992) "Designing emergent behaviors: From local interactions to collective intelligence," in Proc. Int. Conf. Simulation of Adaptive Behavior: From Animals to Animats 2, pp. 432/441

10  Mataric, M. (1992) "Minimizing complexity in controlling a mobile robot population," in Proc. 1992 IEEE Int. Conf. Robot. Automat., Nice, France, pp. 830/835

11  Parker, L. E. (1994) "Heterogeneous Multi-Robot Cooperation," Ph.D. dissertation, Dept. Electr. Eng. Comput. Sci., Mass. Inst. of Technol., Cambridge, MA

12  Yoshida, E., T. Arai, J. Ota, and T. Miki (1994) "Effect of grouping in local communication system of multiple mobile robots," in Proc. 1994 IEEE Int. Conf. Intell. Robots Syst., Munich, Germany, pp. 808/815

13  Yamaguchi, H. (1997) "Adaptive formation control for distributed autonomous mobile robot groups," in Proc. 1997 IEEE Conf. Robot. Automat., Albuquerque, NM

14  Gage, D. W. (1992) "Command control for many-robot systems," Unmanned Syst. Mag., vol. 10, no. 4, pp. 28/34

15  Parker, L. (1993) "Designing control laws for cooperative agent teams," in Proc. 1993 IEEE Int. Conf. Robot. Automat., pp. 582/587

16  Brooks, R. (1986) "A robust layered control system for a mobile robot," IEEE J. Robot. Automat., vol. RA-2, p. 14

17  Balch, T. and R. C. Arkin (1998) "Behavior-Based Formation Control for Multirobot Teams," IEEE Transactions on Robotics and Automation, vol. 14, no. 6

18  Aoyagi, M. and A. Namatame (2005) "Massive Multi-Agent Simulation in 3D", Soft Computing as Transdisciplinary Science and Technology Proceedings of the 4th IEEE International Workshop (WSTST'05), pp. 295/305

# Emergence and Software development Based on a Survey of Emergence Definitions

Joris Deguet[12], Laurent Magnin[2], and Yves Demazeau[1]

[1] Laboratoire Leibniz, 46 avenue Félix Viallet, 38031 Grenoble Cedex, France
joris.deguet@imag.fr, yves.demazeau@imag.fr,
[2] DIRO, Université de Montréal, Montréal, (Québec) H3C 3J7, Canada
magnin@iro.umontreal.ca

## 1 Introduction

Emergence, a concept that first appeared in philosophy [18, 17], has been widely explored in the domains of Multi-agent Systems (MAS) and Complex Systems [25, 13, 14, 5, 4, 6, 2, 12] and is sometimes considered to be the "key ingredient that makes complex systems complex" [24].

On January 30th 2006, we made a one-keyword query for "emergence" papers on computer science specific engines and generalist scientific engines. We retrieved impressive amounts of documents:

**Table 1**

| Search Engine | Number of results |
|---|---|
| ACM | 1606 |
| IEEE | 783 |
| CiteSeer | 8596 |
| ScholarGoogle | 675000 |

However, there is still a lack of well defined Emergence Based Engineering (EBE) methodologies. Before building such methodologies, we have to look at what implies emergence into software development. Since there is multiple definitions of emergence, we will build our study based on five papers[3] that match the following criteria:

- Emergence definition is the primary goal
- It contains a significantly different (and possibly contradictory) approach from other selected papers

---

[3] A previous paper [8] presents a deeper analysis of those emergence definitions.

J. Deguet et al.: *Emergence and Software development Based on a Survey of Emergence Definitions,*
Studies in Computational Intelligence (SCI) **56**,13–21 (2007)
www.springerlink.com                    © Springer-Verlag Berlin Heidelberg 2007

## 2 Emergence definitions, usual software development and Emergence Based Engineering

Emergence Based Engineering can be defined as "efficient methodologies to build systems that will produce emergent (and useful) phenomena". Conceiving Emergence Based Engineering approaches might not be an easy task, so before going further, let see if we could apply the usual software development methodologies to achieve that goal.

In general, how to design a system that will produce phenomena? In fact, as software analysts and developers, we do not want to produce random phenomena (*i.e.* random system behaviours) in general, but a specific set of well defined phenomena specified through requirements. To obtain such phenomena, we know, thanks to our understanding of how a computer handle a code, that we have to design and code the system in a specific way. In other words, we (need to) understand the causality between the code and the results of its execution.

So, is it possible to use traditional design and coding approaches for Emergence Based Engineering? We will in the following sections study that question through definitions of emergence.

### 2.1 Detection and emergence

A first definition of emergence is provided by Bonabeau and Dessalles [4]. Given the two following notions:

detector            defined as "any device which gives a binary response to its input"

relative complexity $C(S|D,T)$ of a system S "where $D$ is a set of detectors and $T$ a set of available tools that allow to compute a description of structures detected through D" which corresponds to the difficulty to describe the system given $T$ and $D$.

Emergence happens when between time $t$ and $t + \Delta t$, two events happen:

1. a detector $D_k$ becomes activated
2. $C_{t+\Delta t}(S|T, D_1, \ldots, D_{k-1}, D_k) < C_t(S|T, D_1, \ldots, D_{k-1})$

This property is likely to happen in a hierarchy of detectors when an upper level entity summarizes states of a lower level. Thus emergence is the apparition of a synthetic entity.

One widely shared feature of emergence definitions is the existence of levels. Bonabeau and Dessalles do not assume levels *a priori* in the definition but suggest that this is a condition *sine qua non* for the complexity discontinuity to happen.

No assumption is made about the system under detection, therefore one can apply this criterion on both artificial and natural systems as long as detection is possible.

**Emergence Based Engineering's Implications**

Usual software development assumes that the expected phenomena produced by the software under development can be predicted directly from the code; in order to design and write that code. Here, we cannot predict the nature of the phenomena statically, but by detectors that can be used only after coding of the system, at runtime... Therefore, usual software development is not suited to design systems that produce emergent phenomena as defined by Bonabeau and Dessalles.

## 2.2 The emergence test

The first definition focused on an observer modelled by a detection apparatus made emergence somehow "subjective" as the complexity measure depends on this apparatus. However, once the observer is defined, emergence only depends on the perceived behaviour. The emergence test [22, 21] introduces the consideration of the system's design in addition to its behaviour.

This *emergence* test involves a system designer and an observer (possibly the same person). Then if the following three conditions hold, the emergence tag is conferred:

Design        The system has been constructed by the designer by describing *local* elementary interactions between components in a language $L_1$

Observation   The observer is *fully aware* of the design, but describes *global* behaviour and properties over a period of time, using a language $L_2$

Surprise      The language of design $L_1$ and the language of observation $L_2$ are distinct, and the causal link between the elementary interactions programmed in $L_1$ and the behaviours observed in $L_2$ is *non-obvious* to the observer, who therefore experiences surprise.

We can consider Bonabeau and Dessalles' $D$ and $T$ as words and syntax of an observation language $L_2$.

The introduction of the design language $L_1$ has two important consequences:

1. Emergence happens between the design and the observation. This defines a *design-to-behaviour emergence*.
2. Existence of $L_1$ restricts the application of this criterion to artificial systems.

Emergence happens when observation and design appear loosely coupled to the observer.

In the field of decentralized artificial intelligence, Demazeau and Müller [11] have made a similar distinction between *internal* and *external* descriptions of agents where internal description refers to the real architecture of an agent and external description refers to its externally perceived behaviour.

**Emergence Based Engineering's Implications**

Classical software engineering requires that the designer of a system coded in language $L_1$ can predict the future behaviour of its system, described by language $L_2$ (used to express the requirements of the system). This is in contradiction with the emergence test as $L_2$ is not likely to be both predictable and surprising. Also, since "emergence happens between the design and the observation", it is *de facto* not possible to conceive an emergent phenomenon during the design phase... Then again, classical software development is not suited to design emergent behaviours.

### 2.3 Simulation emergence

Making the parallel between intelligence and emergence as subjective notions defined by tests can lead to controversy. One answer could be to consider that emergence happens when a large number of scientists agree that it does. Another answer is to make the definition objective. Simulation emergence is such an attempt, focused on the simulation domain.

In Darley [7] we find this definition:

"A true emergent phenomenon is one for which the optimal means of prediction is simulation."

The author defines two means of prediction depending on $n$ the size of a system:

- $s(n)$: the optimal "amount of computation required to simulate a system, and arrive at a prediction of the given phenomenon".
- $u(n)$: stands for "deeper level of understanding", the way we try to avoid computation by "a creative analysis", $u(n)$ is the amount of computation required by this method.

Then the system will be considered as emergent iff $u(n) \geq s(n)$ i.e. direct simulation is optimal relative to the "amount of computation" measure.

The key issue is to understand what a simulation is. Among all the ways to derive the phenomenon in a computable manner, some are simulations, others are "shortcuts". Then optimality of simulation is equivalent to the absence of "shortcuts".

An interesting point is that both authors address the question of emergence's decidability:

- In Bedau's formulation: "One might worry that the concept of weak emergence is fairly useless since we generally have no proof that a given macrostate of a given system is underivable without simulation."
- With Darley's words "Can we determine, for a given system, whether or not it is emergent ?".

If we reformulate as "the global behavior is optimally obtained by running a system made of interacting micro agents", it provides a natural way to apply the definition to multi-agent based simulations.

**Emergence Based Engineering's Implications**

As already said in the "Detection and emergence" section, usual software development assumes that the expected phenomena produced by the software under development can be predicted directly from the code. Here again, since "a true emergent phenomenon is one for which the optimal means of prediction is simulation", the prediction of the behaviour is not possible directly from the code (or at a too high cost), which is in contradiction with usual software development.

### 2.4 Downward causation and emergence

Bedau has defined *weak* emergence with respect to the *strong* emergence based on *downward causation*. This view is illustrated by Timothy O'Connor [20]:

> "to capture a very strong sense in which an emergent's causal influence is irreducible to that of the micro-properties on which it supervenes; it bears its influence in a direct *downward* fashion, in contrast to the operation of a simple structural macro-property, whose causal influence occurs via the activity of the micro-properties which constitutes it."

In order to achieve *downward causation*, Sawyer [23] proposes that:

1. "as in blackboard systems, the emergent frame must be represented as a data structure external to all of the participating agents"
2. "all emergent collective structures must be internalized by each agent, resulting in an agent-internal version of the emergent."
3. "This internalization process is not deterministic and can result in each agent having a slightly different representation."

We believe that $L_1$ and $L_2$ are of significant interest to clarify this issue. It sounds natural to us to consider that everything with causal powers in an artificial system lies in the $L_1$ design language as it must live within algorithm. Thus even if a data structure exists out of the agents at a macro level, it belongs to the design language. Then $L_2$ to $L_1$ causal power is impossible.

Until here we might have mixed design/observation with micro/macro as it is often the same: We conceive agents and we are very happy to show their collective behavior to colleagues. However, it can be interesting to distinguish the micro/macro from design/observation.

Sawyer's definition is based on the existence of a macro entity external to micro agents. This existence might provide causal powers to this entity on agents. Therefore it allows a *macro to micro causation* we can consider as *downward* as scale decreases. However, this is different from O'Connor's view as agents do not constitute the macro entity.

**Emergence Based Engineering's Implications**

"Downward causation" applied to code and behaviour means that the code/algorithm is determined by the behaviour, not the programmer/designer. In others words, we give to the "machine" a description of the expected behaviour and we get some code... That is not usual software design process (again), but kind of machine learning.

### 2.5 Grammar emergence

This last definition of emergence is specific as its scope is limited to systems expressed in a particular grammar model. This model provides intuitive definitions for micro/macro and design/observation distinctions.

Kubik [16] has proposed an approach based on "the whole is more than the sum of its parts" as inspiration and grammars as a modelling tool.

The key idea is to define a "whole" language and a "sum of the parts" language. From an initial configuration, a language is obtained by rewriting using production rules. For a given set of rules $P_i$, the corresponding language is noted $L(P_i)$.

We can sum up the proposal as follows:

$$\underbrace{L(\bigcup_i P_i)}_{Whole} \underbrace{\supset}_{More} \underbrace{superimposition_i}_{Sum} \underbrace{(L(P_i))}_{Parts}$$

We do not give the definition of the *superimposition* operator here.

Emergence is the case of a configuration being in the whole language but not in the sum of parts. The first is obtained by putting all parts together and deriving configurations, the last by deriving configuration for every part separately and putting results together afterward. Putting together is the way we get a macro entity from micro ones, and derivation is the way to get the language ($L_2$) we observe from the rules ($L_1$) we designed.

When someone hears "the whole is more than the sum of its parts", he or she might reply very fast that a system *is* composed of its parts and therefore cannot be more. To go beyond this triviality, Kubik's elegant idea is to switch micro/macro with design/observation. This makes things comparable as Kubik defines his gap between two set of configurations (similar to $L_2$ and a $L_2'$), at the observation level.

Kubik's idea is close to an informal definition of emergence from [10] stated in the VOWELS framework [9] for multi-agent systems (MAS). This framework suggests a description of such systems as agents (A) in their environment (E), using interactions (I) forming an organization (O). Then the pseudo equation from [10]:

$$MAS = A + E + I + O + Emergence$$

can be seen as:

$$\underbrace{L(MAS)}_{Whole} \underbrace{\supset}_{More} \underbrace{\sum_{v \in vowels}}_{Sum} \underbrace{(L(v))}_{Parts}$$

with VOWELS as an alternate micro partition of a macro MAS.

## Emergence Based Engineering's Implications

"The whole is more than the sum of its parts." Usually the design of a complex system is based on its decomposition by the designer to decrease the relative complexity of its subparts. However, it seems that based on that emergence definition when we decompose the system we add a new global complexity that is bigger than what we get by designing the subparts.

Also, usual software development is strongly based on incremental testing which consist in testing first small parts of code (unit testing), then larger parts, then the complete integration of modules. Here, the general behaviour of the system will radically be different than the behaviour of its parts, invalidating their individual validations. So, again, usual software development, which is heavily based on decomposition, is not well suited for the kind of "emergent" system we would like to produce.

# 3 What are the Emergence Based Engineering alternatives?

Based on what we saw in the last chapter, it is not possible to achieve "classical" emergent programming. To summarize that chapter, when a software developer design and code a system that produces phenomena, if she understands how such phenomena will be produced they cannot be qualified as emergent. At the opposite if she does not know what she achieves, she also does not know what behaviour her program will exhibit... In other words, lack of causality understanding, which is recurrent in emergence definitions, is in opposition with usual programming methodologies.

So, to design a system that will produce a given but also emergent phenomenon, we have to employ different methodologies than usual software development. The main idea is to implement or generate the system without knowing "how it works". We can achieve that goal by at least three approaches:

- By imitating phenomena usually considered as emergent. For example, by implementing the mechanisms of an "ants foraging like algorithm" we could expect the same global behaviour for our system, without having to understand why that behaviour appears;

- By using an incremental design process. First step, we implement a generic system that will produce a behaviour. Based on such behaviour, we try to adapt the system to make it producing a behaviour that will be closer to the behaviour we ultimately expect. Then, we analyse again the produced behaviour, try to adapt the system, etc. ;
- By developing self-adaptive systems. In that case, we could understand how the (meta-)system will be able to modify itself to generate new behaviours when the context changes, but we cannot know in advance what solutions it will produce.

## 4 Conclusion and perspectives

We have seen that Emergence Based Engineering needs new software development approaches. To do so, we suggest to use 1) insights provided by definitions and mechanisms suggested by widely accepted emergence examples (social animals, markets), or 2) machine learning techniques (off-line or embedded).

However, still then remains a lot of issues. How to apply those approaches so to generate in fact emergent phenomena? What will be the differences between "non emergent" machine learning and "emergent" machine learning?

But before going further, we need an unified and computable definition of "computer science emergence" to validate the "emergenceness" of such methodologies. In [8], we have isolated a minimal setting, small as definitions are significantly different.

By going through the definitions [8], we have noticed that emphasis is usually put on the criterion proposed. However, for a computational definition, we think the following points should be refined:

- How do we apply levels on existing systems?
- Can we tag a phenomenon as emergent in a computable way?

We might also explore to what extent a specific definition of emergence is linked with definitions of self-organization or complexity and other terms we usually meet in the field of complex agent networks.

Nonetheless, the reason we wanted a Emergence Based Engineering is the "much from little" idea that Holland has associated to emergence [15]. Indeed, since software systems, in particular multi-agent systems, are going bigger and more complex, reducing the size or the complexity of what is needed to build those systems will become more and more essential. Therefore, Emergence Based Engineering (EBE) sounds like an appealing research track.

## 5 Aknowledgements

We thank the French Government's Direction Générale de l'Armement (DGA) for supporting the research reported in this paper.

# References

1. P. Angeline. *Advances in Genetic Programming*, chapter Genetic Programming and Emergent Intelligence. MIT Press, 1994.
2. A. Barabasi and R. Albert. Emergence of scaling in random networks. *Science*, 1999.
3. C. Bernon, M. Gleizes, S. Peyruqueou, and G. Picard. Adelfe, a methodology for adaptive multi-agents systems engineering. In *ESAW 2002*, 2002.
4. E. Bonabeau and J. Dessalles. Detection and emergence. *Intellectica*, 2(25), 1997.
5. E. Bonabeau, J. Dessalles, and A. Grumbach. Characterizing emergent phenomena. *Revue Internationale de Systémique*, 1995.
6. E. Bonabeau, G. Theraulaz, J. Deneubourg, N. Franks, O. Rafelsberger, J. Joly, and S. Blanco. A model for the emergence of pillars, walls and royal chambers in termite nests. *Philosophical Transactions: Biological Sciences*, 1998.
7. V. Darley. Emergent phenomena and complexity. In R. Brooks and P. Maes, editors, *Artificial Life 4*, pages 411–416, 1994.
8. J. Deguet, Y. Demazeau, and L. Magnin. Elements about the emergence issue, a survey of emergence definitions. *ComPlexUs, International Journal on Modelling in Systems Biology, Social, Cognitive and Information Sciences*, 2006.
9. Y. Demazeau. Steps towards multi-agent oriented programming. In *1st International Workshop on Multi Agent Systems*, 1997.
10. Y. Demazeau. Voyelles. Technical report, CNRS, 2001.
11. Y. Demazeau and J. Muller. From reactive to intentional agents. In *Decentralized Artificial Intelligence 2*. Elsevier, 1991.
12. J. Deneubourg, A. Lioni, and C. Detrain. Dynamics of aggregation and emergence of cooperation. *Biol. Bulletin*, 2002.
13. J. Deneubourg, G. Theraulaz, and R. Beckers. Swarm-made architectures. In *Toward a practice of autonomous systems*, 1992.
14. N. Gilbert. Emergence in social simulation. In *Artificial societies: The computer simulation of social life*. UCL Press, 1995.
15. J. Holland. *Emergence: From Chaos to Order*. Perseus Books, 1997.
16. A. Kubik. Toward a formalization of emergence. *Artificial Life*, 9, 2003.
17. G. Lewes. *Problems of Life and Mind*. Trubner and Company, 1874.
18. J. Mill. *System of Logic*. John W. Parker, 1843.
19. J. Muller. Emergence of collective behaviour and problem solving. In *ESAW*, 2003.
20. T. O'Connor. Emergent properties. *American Philosophical Quaterly*, 1994.
21. E. Ronald and M. Sipper. Surprise versus unsurprise: Implications of emergence in robotics. *Robotics and Autonomous Systems*, 2001.
22. E. Ronald, M. Sipper, and M. Capcarrère. Design, observation, surprise! a test of emergence. In *Artificial Life 5*, pages 225–239, 1999.
23. R. Sawyer. Simulating emergence and downward causation in small groups. In *MABS*, 2001.
24. R. Standish. On complexity and emergence. *Complexity International*, 2001.
25. L. Steels. Towards a theory of emergent functionality. In *From Animals to Animats SAB*, 1992.

# The Impact of Network Model on Performance of Load-balancing

Kensuke Fukuda[1], Toshio Hirotsu[2], Satoshi Kurihara[3], Osamu Akashi[4],
Shin-ya Sato[4], and Toshiharu Sugawara[4]

[1] National Institute of Informatics, Tokyo, 101-8432, Japan
[2] Toyohashi University of Technology, Toyohashi, 441-8580, Japan
[3] Osaka University, Toyonaka, 567-0047, Japan
[4] Nippon Telegprah and Telephone Corp., Tokyo 180-8585, Japan

**Summary.** We discuss the applicability of the degree of an agent–the number of links among neighboring agents– to load-balancing for the agent selection and deployment problem. We first introduce agent deployment algorithm that is useful in the design of MAS for load-balancing. Then we propose an agent selection algorithm, which suggests the strategy for selecting appropriate agents to collaborate with. This algorithm only requires the degree of a server agent and the degrees of the node neighboring the server agent, without global information about network structure. Through simulation of several topologies reproduced by the theoretical network models, we show that the use of the local topological information significantly improves the fairness of the servers.

## 1 Introduction

An entity providing variety of services in the Internet can be viewed as an autonomous agent, that behaves autonomously and cooperatively in order to achieve its goals, which depend on the type of service [6, 13]. In future, there may be thousands of agents playing their own different roles on the Internet. Each agent intends to request and receive a necessary service cooperatively from other agents that can finish the service in a shorter processing time or within the deadline. However, it is generally a non-trivial task to achieve better performance of agents. Each agent must determine its behavior from local information and through local interactions among agents to compensate for the lack of a global view of the world. Besides the efficiency of individual agents, the performance of the collective and heterogeneous agents is also a great concern in a mult-agent system (MAS).

Recently, mainly from the statistical physics field, the structure of the Internet has been characterized as being *scale-free* [4], meaning that most nodes have a few links to others, while a few nodes have almost all of the links. Mathematically, the distribution of the degree $k$ of a node for such a network conforms to a power law

$$P(k) \sim k^{-\gamma}, \tag{1}$$

K. Fukuda et al.: *The Impact of Network Model on Performance of Load-balancing*, Studies in Computational Intelligence (SCI) **56**, 23–37 (2007)
www.springerlink.com

where $\gamma$ is the scaling parameter. The Internet topology is known to be scale-free with $\gamma = 2.2$ [10]. The principal implication of this power-law relationship is that the network topology is not only far from *regular* but is also *random*. This scale-free characteristics also appears in many artificial and natural systems, such as the WWW, collaboration networks, metabolic reactions, and food web [5], suggesting that the implications of scale-free network are likely to be useful in many fields. In terms of coordination/cooperation, the performance of an individual agent and the whole MAS is affected by the structure of the network, though each agent has a different role and ability in dynamical communication environment. The structure of the network is environmental information about the world for agents, so an agent must determine its behavior based on a grasp of the environment to some extent. For example, when an agent assigns tasks to other agents, it is often required to select the closest agent in term of the network hop count.

However, in a scale-free network, a hop-count based simple selection strategy may result in both extremely overloaded agents and non-utilized agents because of the unbalanced deployment of agents. Similarly, simple deployment strategies of agents (i.e., random deployment or deployment without considering network structure) assign most agents to the peripheral parts of the network and a few in the center of the network. Consequently, the load of tasks concentrates on a few central agents because of these simple strategies, resulting in the performance degradation of the total MAS and the loss of robustness and scalability.

Therefore, to achieve load-balancing (i.e., *fairness*) requirement without reducing the *efficiency* in MAS, the agents should recognize the topological properties of the network as environmental information, though it cannot have a global view of the network topology. Thus, in this research, we focus on two issues: (1) by deploying agents to appropriate positions in the Internet at the time of design, load-balancing among agents can be achieved by a selection strategy with simple topological information and (2) for already randomly deployed agents in a given topology, agent's selection algorithm based on additional local topological information balances the tasks for agents.

In this paper, we intend to show that the use of local network topology information enables the better load-balancing to be achieved for the agent selection and deployment problem. For this, we evaluate a selection algorithm for client agents, which does require non-local topological information but requires only the degrees of server agents and those of adjacent neighbors, which agents can easily acquire. In particular, focusing on the contribution of the network topology, we perform our simulation in a simplified network model, eliminating other parameters like bandwidth usage and server load, because it is hard to separate an essential mechanism of the network topology in a complex network environment. So, we first investigate the performance of the server selection and deployment algorithms by using the model whose topology contains the scale-free and random network property. From the simulation, we find that the scale-free property is an essential factor to explain the improvement of the fairness for server selection under the real network topology. However, for the comparison between two models that satisfy the scale-free property, we reveal that the proper value of the scope mainly depends on the degree correlation among neighboring nodes rather than the size of the network, meaning that agents does not need to know the systems size, in priori.

## 2 Preliminary

### 2.1 Server Deployment and Selection Problem

First, we describe our target problem. *Server deployment* is the problem of deciding where each server agent (we refer to this as a server) should be deployed in the network. This is a design problem for construction of a MAS, in advance. For a given network topology and server deployment, the *server selection* is a mechanism for deciding how an client agent (hereafter: a client) should select the "best" server among candidates for requesting a task. Both problems are inseparably involved with each other because one cannot assume that a development that is the best for one selection algorithm is the best for all possible selection algorithms, and vice versa. In our simulation, there is one agent on each node in the network; A fixed number of agents are assigned to the role of server, and the rest of the agents are treated as clients. Also, there is one type of task and each client has the same number of tasks.

For the evaluation of the algorithms, we define two types of measures. Efficiency is a measure of whether the cost of the network, which is quantified by the round-trip-time, the distance between agents, or the traffic volume generated, is minimum or not as a whole system. In our simulation, efficiency was defined as the number of hops between a client and a server. The other measure is fairness, which is the measure of how equally the load of the task is balanced among agents. We refer to fairness as the standard deviation of the number of tasks that a server processes, because the size of task to allocate is fixed and the processing power of the servers are the same among them, as described in the next section. A smaller standard deviation corresponds to a better fairness.

### 2.2 Network Topology

We investigate the performance by using two types of network topology: the actual Internet topology and topologies reproduced by theoretical models.

We used the actual Internet topology provided by CAIDA [1], as used in the study [7]. The network consisted of $8,584$ nodes and $19,439$ undirected and unweighted links. A node corresponds to an autonomous system (AS), which indicates a unit of wide-area routing like an ISP, university, or company, and a link corresponds to the data transmission path between two nodes. One advantage of using the AS topology rather than the router-level topology as the network topology is that it provides a coarse-grain view of the Internet, so the algorithms do not need to understand the complex details of the network topology. There is a disadvantage in using the actual network topology though it reflects the real world: one cannot control the network parameters (i.e., statistical properties).

Also, the statistical properties of the AS topology are completely different from those of a random network and a regular network, as characterized by small world and scale-free. Thus, using such a network topology is more realistic and more practical than using artificial (e.g., random) network as done in previous studies, for evaluating the algorithms. In the AS network, the communication cost for an agent related to the efficiency is defined as the number of AS hops from a client to a server.

In order to quantify the key mechanism of the deployment and selection algorithms in detail, we also evaluated the performance of those algorithms under two theoretical network models: the generalized Barabási-Arbert (GBA) model [3] and the connecting nearest-neighbor (CNN) model [12]. The GBA model is based on the concept of preferential attachment, which is a model of WWW behavior that a user tends to link from one's page to a more popular web page. The CNN model is a model for a social network, in which a person tends to have a friendship with the friends of one's friends. The scale-free characteristics is our main interest, and both models can reproduce a scale-free topology (i.e., degree distribution follows a power-law). Moreover, both models can reproduce a random network as well as a scale-free network, if a suitable control parameter is chosen. Thus, we can analyze the performance difference of the algorithm under the network topologies reproduced by two models with the same order of nodes and links as those for the actual network, when the control parameter changes.

## 3 Evaluation Algorithms

### 3.1 Server Deployment Algorithms

We evaluated two types of server deployment algorithms; In *degree-oriented deployment*, a server is assigned to a node in descending order of node degree. Thus, most of the servers have a huge number of neighbors as a result of the power-law relationship. Figure 1 is the cumulative degree distribution in the actual Internet topology. The servers are selected from the right part of the plots in the figure. In *random deployment*, a server is randomly located independent of its degree. In this algorithm, the number of neighbors of selected servers is likely to be very small, because the probability of choosing a bigger value in a power-law distribution is significantly small. Thus, the servers are selected from roughly the left part of the plots. A schematic figure explaining both deployment algorithms also is shown in Figure 2. Note that degree-oriented deployment requires a global view of the network, though random deployment can be determined without any detailed information about network structure.

In the context of MAS, degree-oriented deployment places the servers on more convenient nodes for service from the viewpoint of the network topology. On the other hand, random deployment does not consider topological information, but this does *not* mean uniformly random deployment of the servers on the network topology, i.e., most servers are placed on nodes with small degree $k$ because of the power-law nature. In the simulation, we placed 10 servers on 10 nodes in the network.

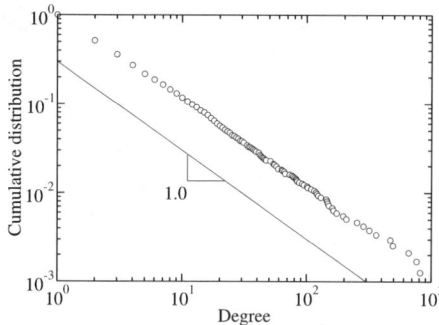

**Fig. 1.** Cumulative degree distribution of the actual Internet topology

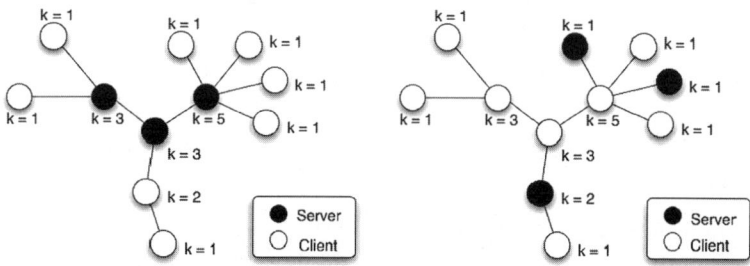

**Fig. 2.** Schematic representation of degree-oriented deployment (left) and random deployment (right)

### 3.2 Server Selection Algorithms

First of all, each client selects the *closest* server, i.e., the minimum AS hop count from the client to the server. Also, a client knows the number of hops from the client to all servers and the degree of servers by means of a routing protocol. If the client finds more than one closest server, it chooses one by using the *reverse-weighted degree (RWD) selection algorithm*. We emphasize that because each client always selects the shortest path, the communication cost of clients in the RWD selection algorithm is minimum so the overall efficiency of agents are maximized.

Suppose that there are $s$ closest servers for a client, and each server $n_i$ ($i = 0 \ldots s - 1$) has a degree $k_i$. In the RWD selection algorithm, a client chooses a specific server from the candidate servers with the following probability:

$$p_i = \frac{k_i^{-\beta}}{\sum_{j=0}^{s-1} k_j^{-\beta}}, \tag{2}$$

where $\beta$ is a weight parameter. For $\beta = 0$, the client selects the server randomly and independently of the degree of a server. For larger $\beta$, the probability of choosing a server with smaller $k_i$ is expected to be higher. The algorithm should provide load-balancing among nodes because it prevents concentration on servers with high degrees. It requires the degree of a server which is non-local information for a client but it is easily available because it can be advertised with the routing information for calculating the shortest path.

As an extension of RWD, we also introduce scope $r$, which is the number of hops from a server. For $r = 0$, the server itself is within the scope. Thus, the clients less than $j$ hops away from the server belong to the scope $r = j$ of the server. The maximum degree of server $n_i$ with scope $r$ is defined as the maximum degree among nodes within $r$ hops from server $n_i$. A server advertises the maximum degree as its own degree $k_i$ in the RWD with scope algorithm. The scope introduces a kind of renormalization of server degrees and has the effect of reducing the sensitivity of fairness due to the smaller-degree server neighboring a huge-degree client node. Figure 3 is a schematic representation of the concept of scope. In this figure, the advertised degrees of the server whose degree is 3 are 5 for $r = 1$ and $r = 2$, though the advertised degree is 3 for $r = 0$.

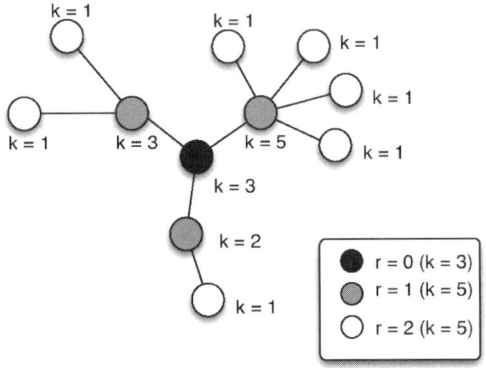

**Fig. 3.** Definition of scope

In summary, the RWD selection algorithm selects the server with the shortest path and lower load average from network topology information, instead of from information obtained by other agents. Similarly, an agent's scope spreads the collection range of information about adjacent neighboring agents, corresponding to the amount of information that the agent can hold. Also note that a client does not need its own location information to select a server. Thus, both server and client can maintain the client states at no cost. We evaluate our algorithm under a simplified network, because the aim is to understand the contribution of the statistical property of network topology. In fact, the algorithm can be extended to use with other environmental information (e.g., bandwidth usage, server load).

# 4 Simulation Results

## 4.1 Actual Topology

We first investigated the performance of degree-oriented and random deployment algorithms for actual Internet topology data as shown in Ref. [7]. Figures 4(a) and (b) demonstrate the fairness improvement for the RWD selection algorithm with scope $r$ under the degree-oriented and random deployments, respectively. Here, the x-axis indicates the weight parameter $\beta$ and the y-axis indicates the standard deviation of the number of tasks that each server served (i.e., a smaller standard deviation shows better fairness). The standard deviation was calculated from 100 different random patterns. Note that $\beta = 0$ and $r = 0$ corresponds to random selection from the candidate servers. Under degree-oriented deployment, larger $\beta$ yielded better fairness

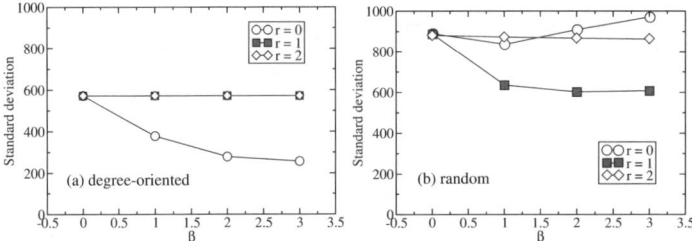

**Fig. 4.** Actual topology

for $r = 0$, showing that the use of degree information improves the total network performance. Nevertheless, for $r > 0$, the standard deviation did not change with increments in $\beta$. Thus, the introduction of scope cancels the advantage of using degree. The reason for this is that the server with the largest $k$ is directly connected to the rest of the servers. Therefore, each client has to choose a server without additional statistical information, that is, selection is the same as random selection.

On the other hand, under random deployment, the fairness became worse for larger $\beta$ without scope, because a server with small $k$ but directly connected to a client with a huge $k$ server had to process more and more tasks as $\beta$ increased; it was even worse than random selection. However, this shortcoming was solved by introducing the scope, i.e., the standard deviation decreased for $r = 1$, though the fairness for $r = 2$ again stayed around the same value as for the random selection. So, a larger scope does not necessarily yield better fairness, and there is a best scope according to the statistical properties of the network.

The details of the performance of these algorithms have been shown in Ref. [7], and the following subsections examine a plausible explanation for the origin of the fairness improvement.

## 4.2 Random or Scale-free

We analyzed two types of basic topology–random networks and scale-free networks–to understand whether scale-free characteristics is really useful information or not. Moreover, in the scale-free networks, we focused on the dependency of the scaling exponent on the network performance. We performed similar simulations to the ones described in the previous subsection by using GBA-generated topologies for degree-oriented deployment and for random deployment as shown in Figs. 5 and 6, respectively. The number of nodes was set to 10000 and the number of links to 20000.

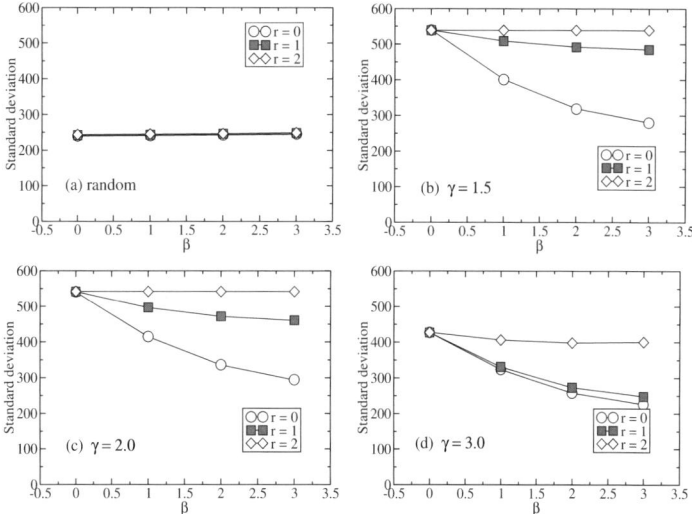

**Fig. 5.** GBA model (n = 10000) for degree-oriented deployment

A panel consists of a different parameter set of the network topology generated by the GBA model: (a) random network, (b) scale-free with $\gamma = 1.5$, (c) scale-free with $\gamma = 2.0$, and (d) scale-free with $\gamma = 3.0$. The scaling exponent of the actual AS network is approximately 2.2, so the behavior for (b) would be expected to resemble that of the actual network. Larger $\gamma$ means that there is a much smaller probability of the appearance of a larger-$k$ node though the plots decay as a power-law.

For degree-oriented deployment (Fig. 5), the result for the random network was different from the others; Neither the use of degree nor scope was affected the performance. This is because all servers had the same degree, so clients had no information about the degree for selecting the server, same as random selection.

For scale free networks, for $r = 0$ the standard deviation decreased for larger $\beta$, meaning that the degree parameter improved the fairness but the scope was negligible. This result is consistent with the result obtained from the actual network topology. In addition, we could confirm that the fairness improve even for $r = 1$ as

$\gamma$ increased. The reason for this is as follows. Although the nine severs had a direct link to the maximum degree server for the actual network topology, there were no such links for the reproduced network topologies.

For random deployment, on the other hand, Fig. 6(a) shows that even for a random network, the performance improved slightly for larger $\beta$. However, the effect of the scope was negligible. Unlike the degree-oriented deployment, the variance of the degree among servers were not zero. This result implies that even a slight difference in degree is useful information for network control.

For the scale-free network, despite the value of $\gamma$, $r = 1$ outperformed the others, though $r = 0$ is closer to the best case for larger $\gamma$. The results for scale-free network are consistent with the result for the actual network, in terms of indicating the usefulness of the degree and the 1-hop scope. In this sense, we can conclude that scale-free characteristics plays a key role in archiving better fairness when the degree information and scope are introduced.

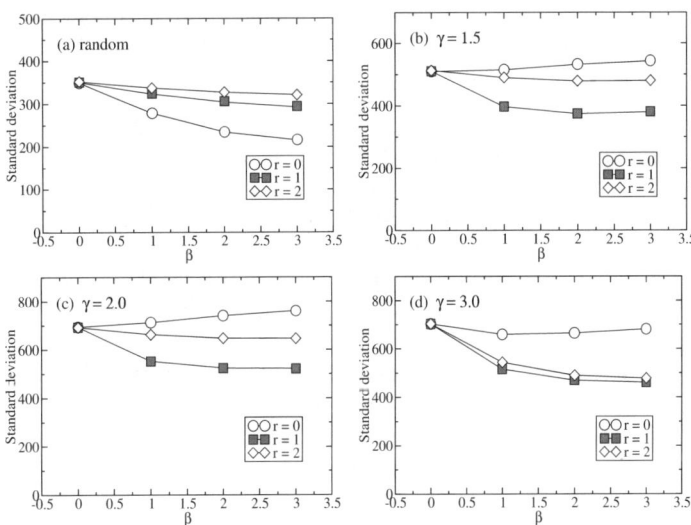

**Fig. 6.** GBA model (n = 10000) for random deployment

## 4.3 Difference in Theoretical Models

### Results for CNN Model

We have discussed the performance of the server deployment and selection problem by using the GBA model. Here, we analyze yet another model, in which a reproduced network is characterized by scale-free property.

Figure 7 indicates the change in fairness for random deployment on CNN generated topologies. The results for the CNN model were remarkably different from

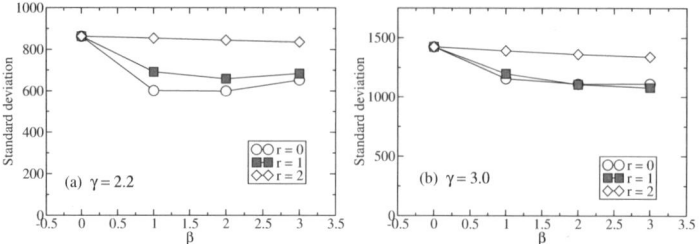

**Fig. 7.** CNN model (n = 10000) for random deployment

those for the GBA model: the fairness curve for $r = 1$ was close to that for $r = 0$, meaning that the introduction of the scope did not improve the fairness in CNN. Furthermore, the standard deviations for $r = 2$ were equal to the value of the random selection. The overall behavior is close to that for degree-oriented deployment on the GBA, rather than that for the random deployment. Thus, besides the scale-free characteristics, some unknown property is also controlling the performance for the RWD scope algorithm in the actual Internet topology.

**Effect of Degree Correlation**

In order to clarify this difference, we focused on the relationship between a node and its neighboring nodes. We quantified this relationship by degree-correlation [10]. The degree-correlation is the slope of the plots between the degree of a node $k$ and the mean number of degrees $< k_{nn} >$ among adjacent neighbors. The value of the slope quantifies the level of the correlation: a positive correlation (i.e., positive value of the slope) means that a larger degree node has neighboring nodes with a larger degree with a higher probability. Thus, a larger-degree node tends to be clustered to larger-degree nodes. On the other hand, if the correlation is negative, a node with larger degree connects to neighbors with smaller degree. The degree of the neighbors is independent of the degree of nodes when the correlation is zero (i.e., flat slope).

   Figure 8 displays the relationship between the degree of a node and the mean degree of neighboring nodes. We can confirm remarkable differences between the topologies generated by the two models. The topology generated by the GBA model had a negative correlation the same as that for the actual Internet topology, particularly the same $\gamma$. However, the topologies generated by CNN were characterized by positive correlations. In addition, they were close to flat for larger $k$. Thus, CNN cannot reproduce the statistics that appear in the actual topology, in terms of degree correlation. In other words, GBA tends to attach a smaller-degree node to a larger-degree node (i.e., to the hub), whereas CNN is inclined to connect a node to those with the same level of degree. It is also noted that the slope for $\gamma = 3.0$ in the CNN is closer to 0, compared with that for $\gamma = 2.0$. This difference might affect the appropriate value of the scope.

   Another method to quantify the relationship between the degree of a node and that of adjacent neighboring nodes is assortativity proposed by Newman [9]. The

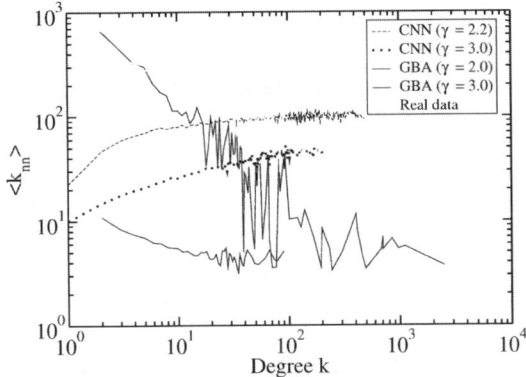

**Fig. 8.** Degree correlation of the network

definition of assortativity is as follows

$$\text{assortativity} \quad \frac{M^{-1} \sum_i j_i k_i - [M^{-1} \sum_i \frac{1}{2}(j_i + k_i)]^2}{M^{-1} \sum_i \frac{1}{2}(j_i^2 + k_i^2) - [M^{-1} \sum_i \frac{1}{2}(j_i + k_i)]^2}, \quad (3)$$

where $j_i, k_i$ are the degrees of the vertices at the ends of the $i$th edge, with $i = 1 \cdots M$. A positive assortativity indicates that a smaller degree node tends to connect to smaller degree nodes, and vice versa, as is known to appear in social networks (e.g., coauthorship and film actor collaborations). Conversely, a negative one indicates that the smaller degree-node attaches to a higher one. Typical examples of a negative assoratity network is the Internet topology and Web. Thus, the assortativity also indicates the connectivity of the nodes as the environmental information.

**Table 1.** Assortativities of networks

| network | assortativity |
|---|---|
| Actual topology | -0.18 |
| BA ($\gamma = 2.0$) | -0.240 |
| BA ($\gamma = 3.0$) | -0.003 |
| CNN ($\gamma = 2.2$) | 0.055 |
| CNN ($\gamma = 3.0$) | 0.167 |

Table 1 represents the assortativities of the topologies used in our simulations. It is also apparent that BA is characterized by the negative assortativity, meaning that a smaller-degree node tends to connect to larger-degree nodes. On the contrary, CNN has a positive assortativity a larger-degree node connected by the same-level-degree nodes, unlike the result for the actual Internet topology.

# 5 Discussion

## 5.1 Effectiveness of Using Topological Information

We discuss the effectiveness of using the topological information (i.e., degree and scope). Our results indicated that the total performance of MAS depends on the statistical properties of the network. The introduction of degree information and the appropriate scope using the actual Internet topology can improve the fairness of servers significantly. Also, the efficiency of the whole system is maximized in a given topology, because the RWD selection algorithm always uses the shortest path from a client to a server.

The effectiveness derived from the server deployment and selection algorithms originated in two topological properties. One is the scale-free characteristics; the relationship between nodes is highly asymmetric. Primarily, strong asymmetry of degree is essential because the client has no information to differentiate the server from candidates without the bias of the degree. For a random network, the asymmetry is smaller than that for the scale-free network, but the use of the degree improves the performance, though the effect was restrictive.

Another property is negative degree correlation to explain the optimality of the scope for random deployment. In the case of random deployment, the server is assigned to a smaller-degree node due to the power-law relationship of the network. Scope is a parameter for checking whether there is a hub node (large-degree node) closer to the server. If the degree correlation is positive, most servers have little chance of finding a hub node because they have smaller degree. In this case, the effect of the scope is negligible. Conversely, for a negative correlation, the servers can find a hub with a low cost, particularly for the case of $\gamma = 2.0$. For example, the degree of a neighboring node of a node with $k = 2$ for $\gamma = 2.0$ in the GBA is about 10 times larger than that $\gamma = 2.2$ for CNN (Fig. 7). Thus, smaller-degree servers are connected to hub nodes in the actual network, so the value of the scope is at most 1. Similarly, when the maximum-degree node is directly connected to smaller-degree servers, the effect of the scope is canceled by the maximum degree. So, the relative distance between smaller-degree server and the maximum-degree node is also important.

Note that the positive degree correlation does not contradict to the scale-free characteristics as we observed in the CNN model. This implies that we should carefully choose the model to generate the network topology, even if a model reproduces scale-free network.

## 5.2 Implication for MAS Design

As the guideline for MAS design, we have shown that a designer must carefully consider the combination of the statistical properties of the network topology, deployment of the agent, and the selection of the agent. The server deployment based on the degree has a flaw when applied to the actual network because of the social constraints (e.g., politics or budget), because it is generally difficult to deploy a server at the hub ISPs. Nevertheless, this deployment showed the best performance for load-balancing in a scale-free network when degree information was used. Moreover, in this scenario, the server does not need degree information of the adjacent neighboring agents; the cost to the agent is minimal.

Next, for random deployment of the server (more realistic scenario), we found that the use of the degree information improved the fairness in both random and scale-free networks. Thus, considering a realistic condition of a network topology ($\gamma = 2$ and negative degree correlation), degree of the neighboring agents can be viewed as relative topological information, so that the statistical information is sufficient parameter to control the network performance. In particular, one important finding in practice is that each server should at least have degree information about the adjacent neighbors ($r = 1$) to achieve better fairness when applying the actual network. Thus, each agent need not to have wider view of the network.

From the viewpoint of scalability, it is also an advantage that the selection algorithm does not need topological information about the client. Therefore, the selection algorithm has scalability against the system size because the state information to be kept by both server and client is minimal. In particular, this property is essential for a massive MAS.However, we need further investigation for scalability of our algorithms in more larger-scale simulations.

Another contribution of our work to the MAS research field is to have shown that the performance of a proposed algorithm in a MAS strongly depends on the topological properties of the underlying network. In particular, even if topologies generated by two models (e.g., inspired by the growth of the web and by the growth of friendship) are scale-free, the best strategy may be different for these two cases. Further studies are needed on the application of statistical information of the underlying network to MAS system, especially to the dynamic formation of collective agents [8].

### 5.3 Application of Our Results to Other Fields

Our simulation setting is simpler than typical simulation settings used by Internet-working research, in order to obtain the essential effect of realizing load-balancing.

So, our findings indicates a possibility to be applied to other systems with the same statistical properties. Currently, we are interested in web applications, because the web structure is characterized by scale-free characteristics ($\gamma \approx 2.0$) and a negative degree correlation. There are some application of scale-free characteristics to searching in the Web or in a peer-to-peer network, which is based on the random walk [2, 11]. However, there are few proposals for using degree and scope as knowledge about the network structure.

Furthermore, we consider that our results are general enough to apply to other research fields, particularly social science. Even in a human network consisting of nodes corresponding to individuals and links between them, the load-balancing of the task among individuals is important issue in the real world. A difference of statistical properties between the Internet and human network is the degree correlation, as shown in the previous section. The structure of the Internet exhibits a negative degree correlation although the CNN model, which models the human network, is characterized by a positive degree correlation. Thus, our simulation results for the CNN model are able to be applied to the problem in the human network, and told us that we should ask a task to a key person with the smaller number of friends though the number of friends for a friend of the key person does not matter. We need further basic studies to understand the applicability of our idea to the human network.

## 6  Conclusion

This paper analyzed on the effectiveness of using statistical parameters of the network structure in a MAS. We demonstrated through simulations that degree and scope are useful parameter for improving the fairness of agent deployment and selection problem in the Internet. In particular, the important implication is that for random deployment on a scale-free network appearing in the actual network, the selection algorithm requires only the degree of the server and its adjacent neighbors. This means that an agent does not have to know the global topological view of the world. Similarly, our simulation results indicated that the use of the degree information is useful in the situation under the CNN topology which models the social network, though the introduction of the scope cancels the performance improvement. Furthermore, we analyzed the key mechanism for achieving the better fairness in the algorithms by comparison with the theoretical models. Consequently, we showed that scale-free characteristics and negative degree correlation play an essential role.

In future work, we will analyze the scalability of our algorithm, and also the applicability of other parameters of the network structure to network control.

## References

1. Cooperative association for internet data analysis (http://www.caida.org/).
2. L.A. Adamic, R.M. Lukose, A.R. Puniyani, and B.A. Huberman, 'Search in power-law networks', *Phys. Rev. E*, **64**, 046135, (2001).
3. R. Albert and A.L. Barabási, 'Statistical mechanics of complex networks', *Rev. Mod. Phys.*, **74**, 47–97, (2002).
4. A.-L. Barabási and R. Albert, 'Emergence of scaling in random networks', *Science*, **286**, 509–512, (1999).
5. S.N. Dorogovtsev and J.F.F. Mendes, *Evolution of Networks: From Biological Nets to the Internet and WWW*, Oxford University Press, Oxford, 2003.
6. E.H. Durfee, V.R. Lesser, and D.D. Corkill, 'Cooperation through communication in a distributed problem solving network', in *Distributed Artifical Intelligence*, ed., M. Huhns, chapter 2, Pitman, (1987).
7. K. Fukuda, S. Sato, O. Akashi, T. Hirotsu, S. Kurihara, and T. Sugawara, 'On the use of hierachical power-law network topology for server selection and allocation in multi-agent systems', in *Proc. of IEEE/WIC/ACM International Conference on Intelligent Agent Technology*, pp. 16–21, (2005).
8. M. E. Gaston and M.desJardins, 'Agent-organized networks for dynamic team formation', *In Proc. of International Joint Conference on Autonomous Agents and Multiagent Systems*, 230–237, (2005).

9.  M.E.J. Newman, 'Assortative mixing in networks', *Phys. Rev. Lett.*, **89**, 208701, (2002).
10. R. Pastor-Satorras, E. Smith, and R.V. Sole, 'Dynamical and correlation properties of the internet', *Phys. Rev. Lett.*, **87**, 258701, (2000).
11. N. Sarshar, P.O. Boykin, and V.P. Roychowdhury, 'Percolation search in power law networks: Making unstructured peer-to-peer networks scalable', *In Proc. of Fourth International Conference on Peer-to-Peer Computing*, 2–9, (2004).
12. A. Vázquez, 'Growing network with local rules: Preferntial attachment, clustering hierarchy, and degree correlations', *Phys. Rev. E*, **67**, 056104, (2003).
13. *Multiagent Systems – A Modern Approach to Distributed Artificial Intelligence*, ed., G. Weiss, MIT Press, (1999).

# Auction-Based Resource Reservation Game in Small World

Zhixing Huang, Yan Tang and Yuhui Qiu

Faculty of Computer and Information Science
Southwest China University {huangzx, ytang, yhqiu}@swu.edu.cn

**Summary.** Due to that the computational resources, such as grid, are not storable, the advance reservation of these resources are necessary for users to request resources from multiple scheduling systems at a specific time. Meanwhile, auction is an appropriate mechanism to reach an economically efficient allocation of goods, services, resources, etc. The auction-based resource reservation models need to be investigated. In this paper, we study the evolutionary dynamics of auction-based resource reservation game played by agents with different bidding strategies. Moreover, our analysis is from the semi-cooperative perspective, which assumes that agents can directly coordinate their bidding with their neighborhoods. We assume that all agents are located in the small world, the user agents can change their strategies through learning the successful bidding experiences from his neighborhoods. Our results indicate that any a part of the bidding strategies will survive in the system which is highly depended on the progress of the competition, and the average satisfaction of the agents could increase after a period of cooperation.

## 1 Introduction

Grid Computing and Peer-to-Peer Computing promises a flexible infrastructure for complex, dynamic and distributed resource sharing. In recent years, distributed computational economy has been recognized as an effective metaphor [3] for the management of Grid resources, it enables the regulation of supply and demand for resources, provides economic incentive for grid service providers, and motivates the grid service consumers to trade off between deadline, budget, and the required level of quality-of-service. Moreover, the Grid is a highly dynamic environment with servers coming on-line, going off-line, and with continuously varying demands from the clients [13]. In most Grid scheduling systems, submitted jobs are initially placed into a queue if there are no available resources. Therefore, there is no guarantee as to when these jobs will be executed. This causes problems in parallel applications, where most of them have dependencies among each other. And grid resources

Z. Huang et al.: *Auction-Based Resource Reservation Game in Small World,* Studies in Computational Intelligence (SCI) **56,** 39–51 (2007)
www.springerlink.com

are not storable, thus that capacity not used today can be put aside for future use. So, the advance reservation of these resources are needed for users to request resources from multiple scheduling systems at a specific time and thus gain simultaneous access to enough resources for their application [4]. As is defined in GRAAP-WG[1], an Advance Reservation is a possibly limited or restricted delegation of a particular resource capability over a demand time interval, obtained by the requestor from the resource owner through a negotiation process [13]. Example resource capabilities: number of processors, amount of memory, disk space, software licences, network bandwidth, etc. Moreover, the works of advance reservations, started within the GGF (Global Grid Forum) to support advance reservations, are currently being added to some test-bed of grid toolkit.

Auction-based approach and posted price-based approach are two most widely used mechanisms in economic-based resource reservation. Auctions provide useful mechanisms for resource allocation problems with automates and self-interested agents [17]. Typical applications include task assignment and distributed scheduling problems, and are characterized with distributed information about agents' local problems and multiple conflicting goals. However, compared with the price-based approach, the adaptive bidding strategies need to be fully analyzed. In [1], A. Byde showed that Genetic Algorithm(GA) can be used to successfully evolve bidding strategies for different auction contexts. M. He [6] proposed a heuristic fuzzy rules and fuzzy reasoning mechanisms in order to make the best bid given the state of the marketplace. K. M. Sim [14] presents a general foundations of designing market-driven strategies of agents. Market-driven agents in his research are guided by four mathematical functions of eagerness, remaining trading time, trading opportunity and competition. These novel strategies outperforms other most prominent algorithms previously developed for auctions in non-cooperative perspective. However, it is hard to find the optimal strategies in Market-based scheduling systems. D. M. Reeves [12] illustrated the difficulty of drawing conclusions about strategy choices in even a relatively simple simultaneous ascending auction game, and J. K. Mason [8] also drew conclusions that advocates of any particular strategy for simultaneous interdependent auction problems are hard to find, as no bidding policy is known to perform generally well, after he had investigated several price-predicting strategies.

In [7], we compared different bidding strategies in auction-based resource reservation games from the standard noncooperative perspective, which assumed that agents did not know their user agents' private information and do not directly coordinate their bidding. In this article, we will extend our previous study with cognitive agents in semi-cooperative contexts, in which the user agents can automatically adapt to their environment and can exchange

---

[1] Advance Reservation State of the Art: Grid Resource Allocation Agreement Protocol Working Group, *URL http://www.fz-juelich.de/zam/RD/coop/ggf/graap/graap-wg.html.*

private information and learn experiences from their neighbors. We will study the evolutionary dynamics of the resource reservation game played by agents in the small world network which has be used to represent the relationship of the human society.

The remainder of this paper is structured as follows. In Section 2, we give a brief review of the small world conceptions about our research. In Section 3, the framework of auction-based advance reservation used in this paper are introduced, and the heuristic bidding strategies of this model are discussed. In Section 4, we demonstrate the empirical evaluation of this heuristic algorithm in both cooperative and noncooperative manner. Finally, in Section 5 we present the conclusions and future work.

## 2 Small World

Following an important body of literature in the field of social-psychology and sociometrics, initiated by S. Milgram [10], the "six degrees of the separation" paradigm of a "small-world". In 1998, D. Watts [16] defined the small world network using two characteristic parameters, i.e., *characteristic path length* and *clustering coefficient* .

A graph $G$ consists of vertices and edges. $V = \{1, \cdots, n\}$ is a set of vertices and $g$ is an adjacency matrix of $G$. $g_{i,j}$ for a pair of vertices $i, j \in V$ indicates an edge between $i$ and $j$. If $g_{i,j} \neq 0$ then an edge exits. An edge is absent in case of $g_{i,j} = 0$.

Let $d(i, j, g)$ be a function which give the length of the shortest path between $i$ and $j$. The *characteristic path length* $L$ is the average of the shortest path length between any two vertices on the network. $L$ is precisely expressed as

$$L = \frac{1}{n(n-1)} \sum_{i} \sum_{j \neq i} d(i, j, g) \tag{1}$$

$N^1(i, g) = \{k \in V \mid g_{i,k} \neq 0\}$ is a set of vertices adjacent to $i$, $E(i, g) = \{(j, k) | g_{j,k} \neq 0, k \neq j, j \in N^1(i, g), k \in N^1(i, g)\}$ is a set of the combinations of vertices in $N^1(i, g)$ when an edge exits between them. For simplicity, let $l_i = | N^1(i, g) |$ be the number of edges connected to $i$. The *clustering coefficient* $C$ indicates the extent to which vertices adjacent to any vertex $v$ are adjacent to each other as average and is defined as

$$C = \frac{1}{n} \sum_{i} \frac{|E(i, g)|}{C_i}, C_i = l_i(l_i - 1)/2 \tag{2}$$

Every vertex is connected to its neighborhoods mutually $p$, where $p$ is a randomness parameter. Edges are changed stochastically as $p$ increases.

In case of $p = 0$, it is a regular network, and both $L$ and $C$ are large. On the other side, it is a random network at $p = 1$, and $L$ and $C$ are small. In the middle between these two extremes, it is called small world network, where $L$ is small and $C$ is large.

The small world networks have been found in many areas, and have been applied in analyzing bilateral games [9], the knowledge and innovation diffusion processes [19] and market organization [18], etc.

In this paper, we use small world networks to simulate the influence of the performance of the bidding agent in resource reservation games when they evolve their bidding strategy only through learning the experience from their neighborhood in the network.

# 3 Auction-Based Resource Reservation Model

In this section, we specifically consider the case of bidding in multiple overlapping English auctions and the state of advance reservation is simplified.

## 3.1 The State of Advance Reservation

As be noted in GRAAP-WG, the state of advance reservation has nine different states. For simplicity, in this paper we only consider the following three most important states:

- Requested: A user has requested a set of resources for a reservation. If the reservation accepted, it goes to being booked. Otherwise, it becomes declined.
- Declined: The reservation is not successfully allocated for some reason.
- Booked: A reservation has been made, and will be honored by the scheduler. From here, the reservation can become active.

That is, the booked reservation can not be altered or be cancelled.

## 3.2 Multiple Ascending Auction

Now, we present a new variant of ascending auction protocol which is different from the standard ascending auction [17] for the general continuous resource reservation protocol. In detail, we consider a multiple English auction market, where each auction is given a start and an active bidding lasting time that may overlap with other auctions. We assume that English auctions work according to the following principles:

1) The auctions proceed in **rounds**. Each auction administrate the price of the service. We don't assume that the different auctions' rounds are synchronized, that is, all auctions move from one round to the next simultaneously

2) In each round, after all the user agents propose their bids, the service provider will select the highest bidding agent as the active one, and then it will update its current price. If there are more than one highest bidding, the

service provider randomly selects one as the *active* agent. If the request of the user agent is declined, the agent is *inactive*. If the agent $i$'s bidding to the service provider $j$ and is active in the last $l_j$ round, the agent $i$ will win this auction and the request of resource reservation is booked. The service provider will reset the service price to its' *reserve* price, and update the start time of the next service, and then a new auction **turn** will begin.

3) In each round, the user agent considers bidding if and only if it is not holding an active bid in an auction or it has not already obtained the service. If user agent $i$ is inactive, the user agent will choose the service provider which it will request to and then decide whether or not to propose the next bid in this server. As we note below, each service provider has its reserve price, and the valid bids to the service provider $j$ must be increased by $\lambda_j$.

Though simultaneous ascending auction protocol does not include announcements of bidder identities, the task size of the active agent will be announced by the auctioneer, and the number of user agents remaining in the system can be known by everyone.

## 3.3 Service Provider

Suppose there are $M$ service providers willing to provide computation service, the services of the providers are homogenous. The services are reserved through auctions, and each service provider runs one auction independently.

Suppose each service provider owns a computation pool with the capacity $c_j$, and computation service can be started at time $s_j$. Each service provider has its lowest price $r_j$ for providing service (also call reserve price) and the minimal increase of the price $\lambda_j$ in the auction. If one agent's bid to the service provider $j$ is the highest bid in the last $l_j$ rounds, the agent $i$ will win this auction and the request of resource reservation is booked. As we have mentioned before, the agent's task size and the service's new start time are announced by provider and can be known by all the user agents. We also call the service provider server later on.

Note that there may exist two kinds of pricing methods for the service in auctions. One is the price for executing a job, and the other is the price of one computation unit. In this paper, we adopt the latter.

## 3.4 User Agent

Next we are to describe the user agent. Suppose there are N user agents, each agent with one task to fulfill before the deadline. Furthermore, we assume that agent $i$ has a budget $m_i$, the size of the task of agent $i$ is $t_i$ and the deadline of the task $d_i$. The information of agents' task size, deadline and budget are private information, and the distributions of these information are all unknown to the agents.

Suppose the current price of the service $j$ is $p_j$, so the cost of user agent is: $b_{ij} = t_i p_j / c_j$. And, if the agent $i$'s task running on server $j$ starts at $s_{ij}$, the end time of the task is $e_{ij} = s_{ij} + t_i/c_j$. The utility of agent is related to the cost of fulfilling the task and the task's end time. In this paper, we use a constant elasticity of substitution (CES) function to present user agent's utility $u_{ij}$. If $d_i \geq e_{ij}$ and $m_i \geq b_{ij}$,

$$u_{ij} = A_i(\delta_{i1}(d_i - e_{ij})^{\rho_i} + \delta_{i2}(m_i - b_{ij})^{\rho_i})^{1/\rho_i} \tag{3}$$

otherwise, $u_{ij} = 0$. So the agents must make trade-offs between the task finish time and the cost of the execution. The satisfaction of the user agent is defined as follows:

$$\theta_i = \frac{u_i}{A_i(\delta_{i1}d_i^{\rho_i} + \delta_{i2}m_i^{\rho_i})^{1/\rho_i}} \tag{4}$$

In the next section, we will investigate several bidding strategies to resource advance reservation.

### 3.5 Bidding Strategies

In contrast to the stand-alone English auction, there is no *dominant strategy* that can be exploited in the multiple auction context [2] and there did not exist Nash equilibrium in even in the very simple scheduling problems [17]. So in this auction-based resource reservation model, the heuristic strategies are adapted.

Before we propose the heuristic algorithms, we firstly make a distinction between the bidding strategies. If the agent's strategy can predict the prices of the next $k$ turns of the auctions, we call the strategy level $k$ strategy. And more specifically, if the agent takes into consideration all the future turns of the auctions, we call this strategy level $*$ strategy. If we only consider the current turn of each auction, there were three most significant strategies has been proposed in this literature [2]:

- *Fix Auction* (0) strategy: randomly select at the beginning of a game the auctions in which bids will be placed and then bid in these auctions only. The agent continues bidding in this auction until its' satisfaction is negative, then it will switch to another auction.
- *Greedy* (0) strategy: unless the agent is active in some auction, bid in whichever auction currently has the highest evaluation.
- *Fix Threshold* (0) strategy: assign an evaluation threshold $\theta \in [0,1]$ for satisfaction degree at the beginning of the auctions. Then bid in whichever auction currently has the highest evaluation until the resource is reserved or *theta* is reached. In the latter case, switch to another random auction until all of them close. In this paper, we set this value of parameter $\theta$ to be 0.5.

In [7] we have proposed two heuristic bidding strategies which outperformed in most cases in the comparison with those three level 0 strategies, there are:

- *Greedy* (1) strategy: unless the agent is active in some auction, based on Equation 4, the agent decides to bid in whichever auction currently has the highest evaluation, or decides not to bid if there exists a higher expected evaluation in the next turn than the current highest evaluation. We use the following equation to predicate the closing price of the next $k$ turn in the same server.

$$p_j(t_k) = r_j + \frac{(p_j(t) - r_j) * lg(n(t_k))}{lg(n(t))} \qquad (5)$$

In this equation, $p_j(t)$ is the current price and $n(t)$ is the current number of agent who are participating in the auctions at the current time $t$. Moreover, $p_j(t_k)$ is the predication of the price on server $j$ and $n(t_k)$ is the number of agents remaining in the system at the time of $t_k$, $n(t_k) \in [1, n(t) - 1]$. For example, we can know that if there is only one user agent in the system, the price of the service on server $j$ will fall down to the server's reserve price. But as the number of agents increases, the competition of the service becomes severe, and the price of the service will increase.

- *Greedy* (∗) strategy: If the agent knows the average size of agents' tasks, it can use the following equation to predicate its task end time.

$$e_{ij} = s_j + \frac{t_j}{c_j} + \bar{t} * \left\lfloor \frac{\bar{t} * (n(t_k) - 1)}{\sum c_j} \right\rfloor \qquad (6)$$

For $n(t_k) \in [1, n(t) - 1]$, based on Equation 5 and Equation 6, the agent can predicate the maximum utility by comparing different utilities. Unless the agent is active in some auction, the agent decides to bid in whichever auction currently with the highest evaluation, or decides not to bid if there exists a higher expected evaluation in future turns than the current highest evaluation.

## 4 Experiment Analysis

### 4.1 Experiment Setup

In this section, we show our three experiments, one is from the noncooperative perspective, the others are from the cooperative perspective. Basically our experiments are performed as follows. All of the agents are located on a network. They plays resource reservation games according to the rules we described previously. Each experiment will run at least 100 times.

We assume that all the agent has the same satisfaction function with $\delta_{i1} = 10$, $\delta_{i2} = 1$, $A_i = 1$, $\rho_i = 2$, $\lambda_j = \$5$, $r_j = \$10$, $l_j = 10$, and all the agent has the same task size $t_i = 100$, same deadline $d_i = 10$ and same budget $m_i = \$100$. In addition, the number of the server is 5 and the capacity of each server is 100.

## 4.2 Experiment 1

In this experiments, the total number of agents $N$ is 50 and the number of each type agent is same at the beginning of the games. It means that all the agents can reserve the resource, however their satisfactions will be different based on their task ending time and costs. In this experiment, our analysis is from the noncooperative perspective. We assume that the agent cannot learn experience from their neighborhood, they don't change their bidding strategies in the games. We compare these strategies based on the average satisfaction of the agents. In this scenario, the average satisfactions of the agents with different bidding strategies are demonstrated in Table 1. From Table 1, we can see the last two high level strategies can obtain much higher satisfaction than other three low level strategies.

**Table 1.** The average satisfaction without cooperation

| Strategy | Satisfaction |
|----------|--------------|
| Fixed(0) | 0.4093 |
| Greedy(0) | 0.6654 |
| Threshold(0) | 0.6273 |
| Greedy(1) | 0.7546 |
| Greedy(*) | 0.7779 |

## 4.3 Experiment 2

In this experiment, we assume that all agents are randomly located in the small world network. We investigate the bidding strategies in the small network with $K = 4$, and the value of $p$ is randomly draw from the set of $\{0.0, 0.1, \cdots, 1.0\}$. The bidding strategy changes after all the auctions are finished in every generation. The agents decide to change stochastically according to their accumulate satisfaction $\vartheta$ earned in the generation, $\tau$ is learning rate.

$$\vartheta_i(t) = (1 - \tau) * \theta_i(t) + \tau * \vartheta_i(t - 1) \tag{7}$$

Then a new bidding strategies is to be adopted there instead of the last one with probability 10%; the new bidding strategy is generated as a copy from the agent who earned the highest satisfaction in the neighborhood. Each experiment here continued up to 100 generations.

In all the games, the average satisfaction of the agents increased along the strategies cooperation, and finally reached a steady value in the end. From figure 1, we can see the average satisfaction becomes steady after about 20 generations. Furthermore, an amazing result in this scenario emerges that nearly all the agents learned to bid at the providers' reserve price after a period of cooperation, for all the cases of $\lambda_j \geq 0.5$, all other strategies would

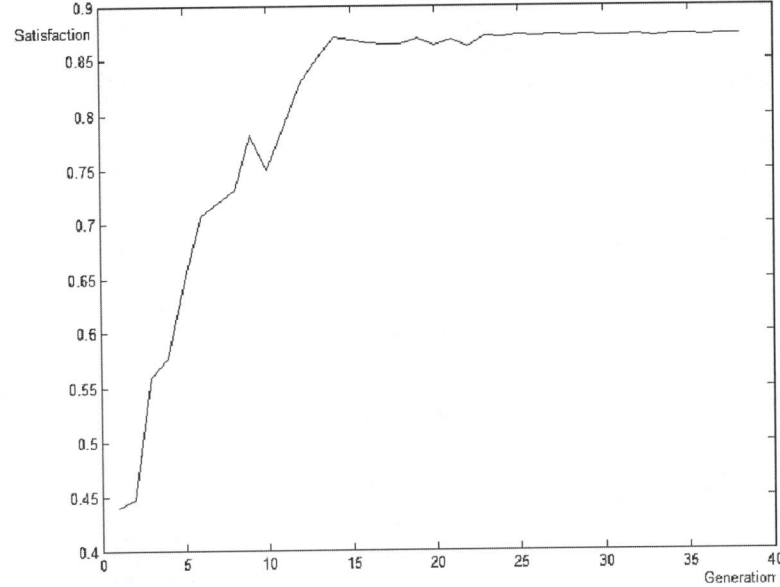

**Fig. 1.** The average satisfaction of agents

converge to $Greedy(*)$, the price for trading the resources were all staying at the reserve price posted by the service providers as seen in Figure 2.

The structure of the small world is one of the factors which influence the converging speed of the strategies. Figure 3 demonstrates the speed change with the variance of $p$. The upper curve shows the situation when all the agents are randomly located in the network, while the lower curve shows the speed at which the agents are sequentially located in the network based on their types.

It should be noted that when $l_j < M$ the bidding strategies will not converge. From Table 2, we can see that also only a part of the strategies will survive as the evolution of the games, the evolutionary result is not only influenced by the strategies in the games, but also dependent on the progress of the games. In addition, some high level strategies will also be replaced by the low level strategies. The average population percentage of the different types are demonstrated in Table 2. Compared with other four strategies, the *Threshold* (0) strategy is dominant in the population and the *Fixed auction*(0) strategy goes to extinct in this games.

Because there exist many simultaneous auctions in the system, the agent with greedy strategy will always find the lowest price auction in the system. When they are changing their bidding from one auction to another auction,

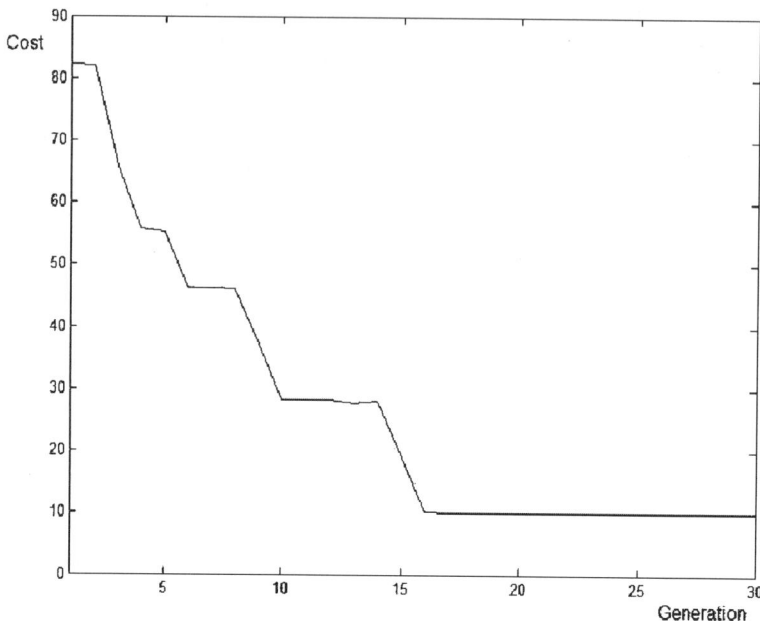

**Fig. 2.** The price converge to reserve price $r_j$

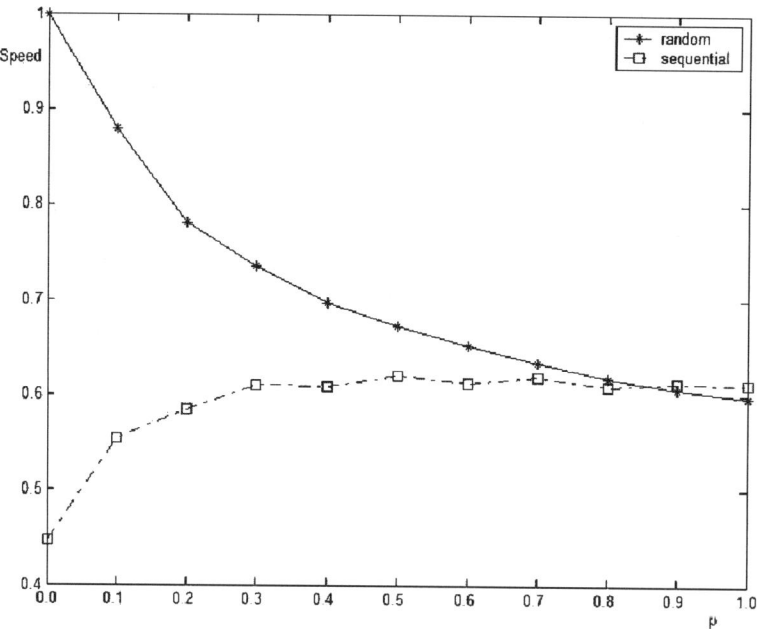

**Fig. 3.** The convergence speed with the variance of $p$

**Table 2.** The proportion of the strategies when $l_j < M$

| Strategy | Average proportion |
|----------|--------------------|
| Fixed(0) | 0.0060 |
| Greedy(0) | 0.2020 |
| Threshold(0) | 0.4300 |
| Greedy(1) | 0.2480 |
| Greedy(*) | 0.1140 |

since $l_j < M$, some auction will be closed at a very lower price at that time. There is no enough time for agent to compete the price of the resource before the auction comes to the closure. Although these results can not reflect the real intention of the strategies, we can still learn from these results when we plan to design the multiple auction protocol and bidding strategies with the consideration of the auction close time.

### 4.4 Experiment 3

We assume that all the agents have the same satisfaction function with $\delta_{i1} = 10$, $\delta_{i2} = 1$, $A_i = 1$, $\rho_i = 2$, $l_j = 10$, $\lambda_j = \$5$, and all the agents have the same task size $t_i = 100$, same deadline $d_i = 10$, same budget $m_i = \$100$. In addition, the number of the server is 5 and the capacity of server is 100, the number of each type agent is the same at the beginning of the games. These parameters are same with experiment 2, except that in this experiment the total number of agents $N$ is 60. It means that only a part of the agents can reserve the resource. In this scenario, the $Fix(0)$ strategy will be dominant, and the agents who reserved the resource will use up all their budgets.

**Table 3.** The proportion of the strategies when $N = 60$

| Strategy | Average proportion |
|----------|--------------------|
| Fixed(0) | 0.5600 |
| Greedy(0) | 0.4400 |
| Threshold(0) | 0.0000 |
| Greedy(1) | 0.0000 |
| Greedy(*) | 0.0000 |

## 5 Conclusion and Future Work

Market-based scheduling to grid scheduling is an exciting research topic. In this paper, we study the evolutionary dynamics of auction-based resource reservation games played by agents with different bidding strategies. Moreover, our analysis is from the semi-cooperative perspective, which assumes that agents can directly coordinate their bidding strategies with their neighborhood. We assume that all agents are located in the small world network, and the user agents can change their strategies by learning successful bidding experiences from his neighborhood. Our results indicate that any part of the bidding strategies will survive in the system which is highly dependent on the progress of the competition, and the average satisfaction of the agents could increase after a period of cooperation.

In current work, we assume that our agents have little prior information about the system, and their bidding strategies are quite myopic. We believe that there exist more effective decision mechanisms which use the explicit probability distribution over possible agent's task size, budget, and the predication of each auction's closing time to reserve resource. So more different price predication functions and bidding strategies need to be investigated both in cooperative and noncooperative environments. Moreover, we will continue to study what factors will influence the proportion of the survival strategies in terms of the cooperation in the small world.

# References

1. Byde A (2001) A comparison among bidding algorithms for multiple auctions. Agent-mediated electronic commerce iv, designing mechanisms and systems, workshop on agent mediated electronic commerce, pp 1-16
2. Byde A, Preist C, Jennings NR (2002) Decision procedures for multiple auctions. In: Proc. of the first int'l joint conference on autonomous agents and multi-agents system, pp 613-620
3. Buyya R, Abramson D, Giddy J, Stockinger H (2002) Economic models for resource management and scheduling in Grid computing, Concurrency and Computation, Vol 14, Issue 13-15, Wiley Press, USA, pp 1507-1542
4. Cheliotis G, Kneyon C, Buyya R (2004) 10 lessons from finance for commercial sharing of it resources, In: Peer-to-Peer Computing: The evolution of a disruptive technology, IRM Press, Hershey, pp 244-264
5. Galstyan A, Czajkowski K, Kristina L (2003) Resource allocation in the grid using reinforcement learning. In: The 3rd int'l joint conference on autonomous agents and multiagent systems, Vol 3, pp 1314-1315
6. He M, Jennings NR, Prugel BA (2004) An adaptive bidding agent for multiple English auctions: A neuro-fuzzy approach, In: Proc. IEEE Conf. on Fuzzy Systems, Budapest, Hungary, pp 1519-1524
7. Huang ZX, Qiu YH (2005) A comparison of advance resource reservation bidding strategies in sequential ascending auctions. In: Proc. the 7th Asia Pacific Web Conference. Springer, pp 742-752
8. Mackie-Mason JK, Osepayshivili A, Reeves DM, Wellman MP (2004) Price prediction strategies for market-based scheduling, In: 18th int'l conference on automated planning and scheduling, pp 244-252
9. Masahiro O, Mitsuru I (2005) Prisoner's dilemma game on network, In: Proc. 8th pacific-rim int'l workshop on multi-agents, pp 9-22

10. Milgram S (1967) The small-world problem, Psychology Today, pp 60-67
11. Phan D (2004) From agent-based computational economics towards cognitive economics, In: Bourgine P, Nadal JP (eds) Cognitive Economics. Springer, Berlin Heidelberg, pp 371-398
12. Reeves DM, Wellman MP, MacKie-Mason JK, Osepayshvili A (2005) Exploring bidding strategies for market-based scheduling, Decision Support Systems 39: 67-85
13. Smith W, Froster I, Taylor V (2000) Scheduling with advanced reservations, In: Proc. of 14th international parallel and distributed processing symposium, pp 127-135
14. Sim KM, Choi CY (2003) Agent that react to changing market situations. IEEE Transations on Systems, Man and Cybernetics, Part B: Cybernetics, Vol 33, No 2, pp 188-201
15. Vytelingum P, Dash RK, David E, Jennings NR (2004) A risk-based bidding strategy for continuous double auctions, In: Proc. 16th european conference on artificial intelligence, Valencia, Spain, pp 79-83
16. Watts DJ, Strogatz SH (1998) Collective dynamics of 'small-world' networks. Nature 393: 440-442
17. Wellman MP, Walsh WE, Wurman PR, Mackie-Mason JK (2001) Auction protocols for decentralized scheduling, Game and Economic Behavior 35: 271-303
18. Wilhite A (2001) Bilateral trade and 'small-world' networks, Computational Economics, Vol 18, No 1, pp 49-64
19. Zimmermann JB (2004) Social networks and economics dynamics, In: Bourgine, Nadel (eds) op.cit. pp 399-418

# Navigational Information as Emergent Intelligence of Spontaneously Structuring Web Space

Takashi Ishikawa

Department of Computer and Information Engineering,
Nippon Institute of Technology,
4-1 Gakuendai, Miyashiro-town,
345-8501 Saitama, Japan
tisikawa@nit.ac.jp

**Abstract.** The paper describes emergent properties of spontaneously structuring web space such as collaborative editing encyclopedias and proposes a method to extract navigational information from the link structure of the web space. The navigational information consists of hyperlinks that are organized with approximate subsumption relations between sets of hyperlinks in the web pages. A case study for Wikipedia and Everything2 exemplifies the effectiveness of the proposed method that allows readers of free online encyclopedias to brows more efficiently.

## 1 Introduction

Web technologies have provided us a new medium of structuring knowledge in a flexible and efficient way. We call the medium hypertext, in which fragments of text are structured using hyperlinks that allow the reader to proceed to related fragments from the specified position of the text. A hyperlink is meta-information embedded in the text without affecting the contents. Whereas usual linear texts are structured with the contents its self. The author of a hypertext is able to restructure the text by changing the hyperlinks embedded in the text, but this flexibility cause the problem that the reader is hard to understand overall structure of the hypertext, that is so called glost in hyperspaceh phenomenon [10]. To overcome the problem authors of hypertexts provide secondary information such as summary, list and index, but they impede the efficiency of reading and disturb the thinking flow of the reader.

Collaborative authoring tools using web technology enable to describe many kinds of contents in the web pages as online encyclopedias [5]. A usual web page of the online encyclopedias is a list of entries or posts each of which

T. Ishikawa: *Navigational Information as Emergent Intelligence of Spontaneously Structuring Web Space*, Studies in Computational Intelligence (SCI) **56**,53–65 (2007)
www.springerlink.com

is a hypertext including hyperlinks to connect to other web pages. The reader of an online encyclopedias selects an entry by the title or a hyperlink in other web page, reads the contents and proceeds to other entries in the same web page or other web pages by clicking an anchor text. The decision where to proceed is made by evaluating relevance of anchor texts to the interest of the reader in reading the entry. Therefore the reader has to read the whole contents of the entry to decide where to proceed. Contrastingly web pages in information sites such as Yahoo [23] have navigational information as category lists or local navigation menus so as to guide the readers in selecting anchor texts without reading the whole contents. The lack of navigational information in online encyclopedias is a deficiency of media to describe a lot of information or knowledge in the web pages.

The research hypothesis of the research is that *there exists implicit navigational in-formation as emergent intelligence of spontaneously structuring web space such as free online encyclopedias.* In the hypothesis, navigational information means category lists or local navigation menus to guide the readers in selecting linked web pages. Looking navigational information is more efficient than reading whole contents in the web page, so it will become more efficient to use online encyclopedias as an information medium. Issues in verifying the hypothesis are that how the navigational information can be generated automatically from the contents and that how to add the navigational information to the web pages of online encyclopedias. An idea to approach the first issue is that the network structure of hyperlinks in web pages reflects the structure of information such as concept hierarchy so that the structural information must be extracted from the link structure [2, 3]. The second issue can be solved by implementing the method for extracting navigational information from link structure of online encyclopedias.

The aim of the research is to develop a computational method to extract navigational information from the link structure of online encyclopedias based on the research hypothesis. The navigational information means a subset of anchor texts in a web page to reduce items to be selected by the readers for the sake of search efficiency. The items in the subset are preferable being structured by the order of importance or relevance and selected so as not to reduce web pages that are reachable through the selected anchor texts. Therefore the problem to be solved is a kind of optimization problem to valance search efficiency and information amount.

## 2 Spontaneously Structuring Web Space

Collaborative free online encyclopedias such as Wikipedia [21] and Everything2 [6] are built with collaborative work in which the authors create and edit each own article with references to other articles. Articles in those encyclopedias are web pages that are created with web-based authoring system and references in the articles are anchor texts with hyperlink to other web pages in the encyclopedia and outside of the web site. An intrinsic property of the encyclopedias is that there is no global structuring scheme that organizes the articles in categorized manner as in usual human edited encyclopedias. So the content of the online encyclopedias forms spontaneously structuring web space in that relationship between articles is made by the collaborative work of independent authors.

While each author of free online encyclopedias does not organize overall structure of the encyclopedias, the link structure of the encyclopedias forms scale-free small-world networks [11, 12]. Out-degree distribution of the network obeys almost power law with exponent nearly 2.5, which means there exists hierarchical structure in the network [4, 19]. Figure 1 shows the out-degree distribution of a fixed reference of Wikipedia on 24 July 2004 [7], where number of node is 299,651 and number of links is 5,637,623. The hierarchical structure seems to be corresponding to the concept hierarchy of topics in the encyclopedia. As in thesauri in information retrieval systems, anchor texts with hyperlinks in the encyclopedia correspond to broader-term (BT), narrower term (NT), and related term (RT). Concept hierarchy is made up with BT and NT relations between terms, which represent class-subclass relation between concepts.

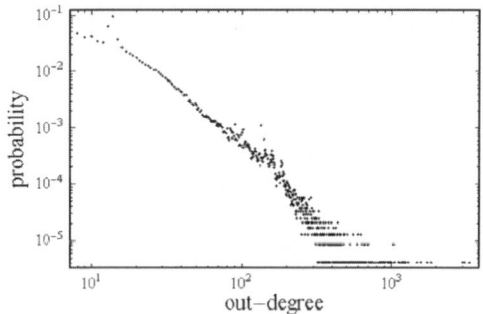

**Fig. 1.** Out-degree distribution of a fixed reference of Wikipedia

The spontaneously structuring property of the free online encyclopedias stems from a special function of the web-based authoring system used for the encyclopedias. That is, if an author creates an article with anchor texts linked to non-existing articles, then the system indicates that articles for the anchor texts must be created by high-lighting the anchor texts automatically. Those anchor texts represent BT, NT, and RT in a topic of the article, so the related articles are gradually created and the concept hierarchy is formed spontaneously. Furthermore the concept hierarchy is formed by the collaborative work of many authors, therefore it represents common understanding of the authors of the encyclopedias. In the result, the concept hierarchy in web space of free online encyclopedias is considered to be emergent intelligence on the network of spontaneously structuring articles.

From the viewpoint of web navigation, a deficiency of free online encyclopedias is that the concept hierarchy in the encyclopedias is implicit so that the user cannot distinguish whether an anchor text is BT or NT. The thesauri in information retrieval systems explicitly describe the relation between terms as BT, NT, and RT. These relations efficiently guide information retrieval in selecting appropriate terms to specify the target information. If the relations between anchor texts in the free online encyclopedias are provided in any way, the deficiency of the encyclopedias removed away and the usefulness of them will increase. In the following sections, we will describe a method to solve the problem.

## 3 Navigational Information in Web Space

There are two methods in using World Wide Web as a information source, one is to use keyword search with search engines and the other is to track hyperlink s in a web page found by the keyword search or by URL directly. In general the keyword search does not output desired information directly, so selection of search results and tracking hyperlinks are needed to find a web page of desired information. To assist users in selecting hyperlink s in web pages, many web sites are designed to have global navigation (i.e., menu) to select an appropriate web page in the whole web site and local navigation (i.e., topics) to select a desired web page linked from a certain web page [17]. We call these navigations *navigational information* in web space. In web space of free online encyclopedias, since there is usually no local navigation in each article, it is hard to find desired information only by tracking hyperlink s. The lack of local navigation in the web space is due to difficulty of organizing topics in a article so as to satisfy needs of all users by a certain organizing criterion. Furthermore since the web space of free online encyclopedias is evolving continuously, it is hard to use a fixed criterion to organize dynamically changing topics.

A clue to cope with the lack of local navigation in free online encyclopedias is that an anchor text in a web page represents a topic or a concept related to a topic of the web page. By a hierarchical nature of concepts, topics of related web pages have relations of BT, NT, and RT as in the thesauri in information retrieval systems. If the relations between topics of related web pages can be determined by properties of the web space, then it is able to construct a local navigation using concept hierarchy of related topics in the web space. Furthermore if relative importance of related topics in a web page can be determined too, then it is able to select important topics in the web page. But how to do it?

By assuming that anchor texts in a web page of free online encyclopedias represent concepts corresponding to BT, NT, and RT as in thesauri, the problem reduces to how to discriminate the types relations between concepts. A key idea of the method in the paper to solve the reduced problem is to use link structure between web pages to discriminate the types of relations.

Hypothesises for the discrimination of relation between concepts in web space using link structure are the following.

1. If web pages are mutually linked, then the conceptual relation between them is BT or NT, and if not, then the relation is RT.
2. If web pages are mutually linked and the number of anchor texts in one page $A$ exceeds that of another page $B$, then the conceptual relation between them is that the page $A$ corresponds to BT to the page $B$.
3. The ratio of the number of common anchor texts in web pages indicates the strength of relation between them (i.e., relative importance with each other).

In the next section, we will describe a method to extract navigational information from link structure of web pages based on the above hypothesisses.

## 4 Extracting Navigational Information from Link Structure

This section describes a computational method to extract navigational information from link structure of web space. The navigational information means a subset of anchor texts in a web page to reduce items to be selected by the readers for the sake of search efficiency. The items in the subset are preferable being structured by the order of importance or relevance and selected so as not to reduce web pages that are reachable through the selected anchor texts. Therefore the problem to be solved is a kind of optimization problem to valance search efficiency and information amount. There are some known methods applicable to construct local navigation as the followings.

1. Tree construction [8]
   a) Shortest path tree
   b) Minimum spanning tree
2. Clustering
   a) Hierarchical clustering by similarity [13]
   b) Community structure [14]
3. Link analysis [9]
   a) k-clique
   b) k-core

Shortest path tree and hierarchical clustering are hard to apply in general to large networks in computational complexity. Community structure is not suitable for extracting local structure of networks since it needs in principle optimization for the whole network. Other methods are not applicable directly to construction of local navigation.

To make browsing in web space more efficient, it needs to reduce items to be selected in each web page without reducing number of web pages that are reachable from a current web page. If web pages constitute a complete graph by hyperlinks between them, then all of the web pages are reachable with each other. But in usual situation some of web pages will not be reachable when a part of web pages are removed from. So the problem becomes to select a subset of web pages linked to a given web page so as to maximize the number of reachable web pages from the given web page for a given reduction ratio of the selected subset. If the ratio of reachable web pages to all linked web pages exceeds the reduction ratio, then abstraction of the network structure is effective in the meaning that the loss of information reduced with suitable selection. So we call the selection process *link abstraction.*

A principle of link abstraction is that if a set of hyperlinks in a web page is a subset of hyperlinks in another web page, then the subsumed web page can be removed from a selected subset of web pages without loss of reachable web pages. But this situation is ideal, so we define subsumption ratio of two sets as follows.

**Definition 1.** *A subsumption ratio SR(A, B) of a set A to a set B is a fraction that the number of common elements in A and B divided by the number of elements in A.*

SR($A$, $B$) ranges 0 to 1, SR($A$, $B$) is 1 when $A$ is subsumed by $B$, and SR($A$, $B$) is less than 1 and greater than 0 when $A$ is partly subsumed by $B$. Subsumption ratio enables to treat approximately relations between web pages from the viewpoint of set theory as shown in Figure 2. To treat hyperlink s in web pages as a set, we assume that a hyperlink is specified uniquely with a referent URL.

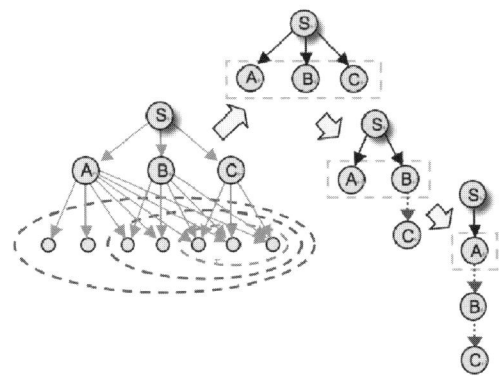

**Fig. 2.** An illustrative example of link abstraction

Link abstraction uses greedy algorithm to select web pages for a given web page as follows.

**Function:** To select a subset of linked web pages in a given web page so as to minimize the loss of reachable web pages

**Input:** a network of web pages and hyperlinks, a current web page, a threshold for subsumption ratio

**Output:** a set of selected hyperlinks in the current web page

**Process:**   1. Compute subsumption ratios for all pairs of web pages in the network.
2. Make a set of web pages that are referred by hyperlinks in the current web page.
3. Sort the set in descendant order of subsumption ratio of the current web page.
4. Remove a web page if it is subsumed from a web page in the set from the tail.
5. Make a set of hyperlinks that refer to the selected web pages.

# 5 Experimental Result

In this section we describe experiments for the proposed method applied to Wikipedia [21] and Everything2 [6] as examples of free online encyclopedias.

## 5.1 A case study of Wikipedia

In order to evaluate the effectiveness of the proposed method for constructing local navigation in an online encyclopedia Wikipedia [21], we have conducted an experiment using a fixed reference of Wikipedia on 24 July 2004 [7]. In the experiment we extracted first all web pages in the site by traversing links from the top page of the site. The average out-degree of the network is 23.5 and number of reachable web pages in the site is 239,838. The details of the experiment are described below.

*Objective* To evaluate the effectiveness of the proposed method for an online encyclopedia Wikipedia.

*Method* Evaluation was done using the following two measures.

1. Reachable ratio of web pages within distance 2 from items in local navigation.
2. Subjective evaluation of importance for the selected items in local navigation.

In evaluating reachable ratio we examined relation between reachable ratio and reduction rate of navigation by varying threshold for subsumption ratio.

*Material* The topics used in the experiment are Scale-free network (as a current topic), Network (as a BT), Link (as RT), and Main Page of the site. Statistics of out-degree and number of linked pages within distance 2 for the selected topics are summarized in Table 1.

**Table 1.** Statistics of web pages used in the experiment

| Topic | Out-degree | Number of linked pages |
|-------|-----------|------------------------|
| Scale-free | 12 | 1,275 |
| Network | 33 | 1,998 |
| Link | 37 | 2,112 |
| Main Page | 99 | 7,177 |

*Result* Figure 3 shows the relation between reachable ratio and reduction
rate of local navigation by varying threshold for subsumption ratio between
sets of web pages. The threshold for subsumption ratio to omit items from
local navigation is varied from 0.01 to 0.5. The result shown in Figure 3
indicates that reachable ratios for all topics exceed reduction rates, therefore
the proposed method is effective in reducing items in local navigation for
the case of Wikipedia. Furthermore the reachable ratio is nearly 0.8 for the
reduction rate 0.5, therefore the reduction effect of the method is satisfiable
for a real application. Figure 4 shows the constructed local navigation for
the topic Network as an example to demonstrate the validity of the method
where the threshold is set to 0.1. Items with dash in the head are items that
are subsumed with the items above them. The hierarchical structures enclosed
with dashed lines show the effectiveness of the proposed method by subjective
evaluation of relevance of the grouped items.

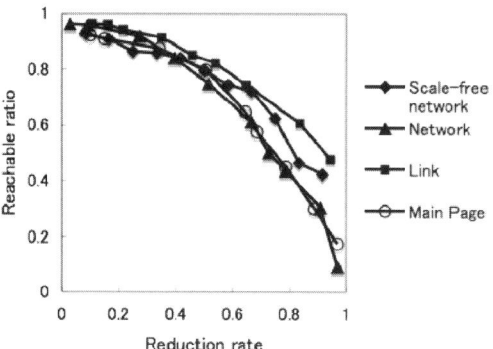

**Fig. 3.** Relation between reachable ratio and reduction rate

## 5.2 A case study of Everything2

We have conducted another experiment to evaluate the effectiveness of the
proposed method for extracting navigational information from link structure

of Everything2 [6]. To apply link analysis to web pages in Everything2, it needs to treat many to many relations between web pages that usually consists of multiple posts concerning different topics in one web page. The problem is to separate each posts in a web page and to determine which post is related to each posts. The problem is hard to solve in general, therefore we address only a specific site Everything2 to demonstrate the effectiveness of the method.

In Everything2 a writeup is the text of an individual user's contribution to a node (i.e., web page) [22]. A user can contribute one writeup to any given node and only the author of a writeup can edit the writeup. There are two types of links in Everything2, one is a soft link that is specified with a title, and the other is hard link that is specified with a node id. A soft link refers to a web page (i.e., node) that consists of multiple writeups while a hard link refers to an individual writeup. A writeup may be displayed in a web page including the writeup or in a web page that consists of the writeup only. In Everything2 these two types of links are used without attention to the distinction of their functions. The situation induces complicated many to many relations between web pages.

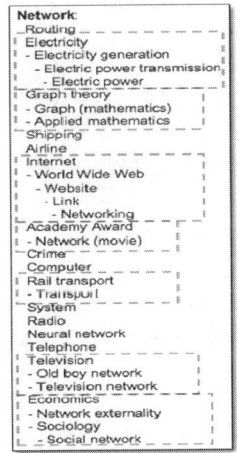

**Fig. 4.** Selected items in a constructed local navigation for Network

In Everything2, when the node designated by a soft link with a node title does not exists, then some titles similar to the title in the soft link are displayed in a Findings page. This function is convenient for human users, but requires an additional function of web crawlers to select the most relevant title from the titles in the Findings page. For the purpose we adopt least difference criterion that select a title with the least numbers of not coincident characters around contiguous coincident characters.

To treat the situation in Everything2, we made the following assumptions:

1. A web page linked with a hard link consists of only one writeup.

2. Writeups in one web page can be separated by syntactical analysis of HTML code.
3. Strength of relation between writeups is proportional to the number of common links.

The assumption 1 and 2 are probably a specification of Everything2 software so that these are correct except for bugs in data. The assumption 3 is essential in writeup selection and grounded on association hypothesis in psychology [18]. These assumptions seem to be valid for web pages used in the experiment described in the previous section. The details of the experiment is described below.

*Objective* To evaluate the effectiveness of the proposed method for an online encyclopedia Everything2.

*Method* Evaluation was done using the following measures.

1. Measures for writeup selection
   **recall:** number of extracted writeups / number of linked writeups
   **precision:** number of correct writeups / number of extracted writeups
   **reduction rate:** number of selected writeups / total number of linked writeups
2. Measures for link abstraction
   **reachable rate:** number of reachable writeups via navigation / number of reachable writeups
   **reduction rate:** number of selected writeups / number of linked writeups
   **grouping relevancy:** subjective evaluation of grouped writeups

The algorithm to select writeups in Everything2 weblog is used for a simple classification based on the assumptions described above.

**Function:** To select writeups in a weblog
**Input:** Set of web pages including multiple writeups
**Output:** Set of individual writeups identified with node ids.

*Material* Everything2 makes it easy for potential authors to contribute, where the content for a node is entered in plain text, which Everything2 converts into HTML. We used 15 sampled nodes of which titles are Blogosphere (place), Network (idea, thing), world wide web (idea, thing), weblog (idea, thing), link (thing), Internet (idea, thing), graph (idea), hypertext (thing, 2 writeups), blog (idea, 2 writeups).

Figure 5 depicts an example result of the writeup selection algorithm applied to a web page with title 'Network' in Everything2.

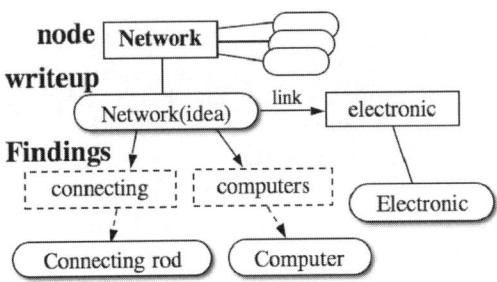

**Fig. 5.** An example result of writeup selection

*Result* Followings are the averaged measurements for the samples n = 15.

1. Measures for writeup selection
   **recall** = 230 / 327 = 0.7034
   **precision** = 200 / 230 = 0.8696
   **reduction rate** = 327 / 989 = 0.3306
2. Measurements for link abstraction
   **reachable rate** = 1510 / 2047 = 0.7381
   **reduction rate** = 154 / 230 = 0.6176
   **grouping relevancy** = GOOD

The following listing shows an example of extracted navigational information for a current node eNetworkf in Everything2. The bold items are selected items and the others are removed items. Numbers in the parentheses for each item are ratio of common items in the current web page.

```
Local Navigation for "Network (thing)
@Everything2.com" : minRatio = 0.1
     Odd (0.0303)
          - even (0.0645)
     Geom (0.0279)
          - exterior (0.125)
          - Sorting vertices for a 3D mesh (0.0344)
          - triangle (0.027)
     point (0.0222)
          - region (0.1052)
     Travel (0.01428)
     Edge (0.0056)
     face (0.0025)
     * reduction rate = 6 / 11 = 0.5454
```

# 6 Discussion

The result of case studies for Wikipedia and Everything2 by the experiments mentioned above demonstrates the effectiveness of the proposed method for extracting navigational information from link structure of web space. Recall and precision for article selection are at satisfiable rate for the real use and also reachable rate exceeds reduction rate in considerable amount. Furthermore grouping relevancy in the selected hyperlinks is judged as valid from subjective evaluation.

# 7 Related Work

Graph-base approach to analyze the network of World Wide Web revealed that it has scale-free structure and small-world property [1, 20]. The scale-free structure implies that there are small numbers of hubs connected to many web pages, which are connected to small number of others. On the other hand, the small-world property means that any pair of web pages are mutually reachable through short path length. These two properties are combined into scale-free small world (SFSW) structure, and on this SFSW, the web page update algorithm performs better than the reinforcement learning algorithm [16]. This is one of the reason why free online encyclopedias has grown rapidly.

Some techniques in the previous research such as building thesaurus [3], adaptive navigation [15], and finding community [14] are applicable to make navigational information with further modification, but no method in literature directly applicable to web space is found to our knowledge.

# 8 Conclusion

The paper has described a method to extract navigational information from link structure in free online encyclopedias such as Wikipedia and Everything2. The navigational information consists of hyperlinks that are organized with approximate subsumption relations between sets of hyperlinks in the web pages. A case studies of Wikipedia and Everything2 exemplifies the effectiveness of the proposed method that allows reader of free online encyclopedias to brows more efficiently. A future work is to implements the method in customized web browser for reading free online encyclopedias.

# References

1. Barabasi, A. (2003) *Linked: How Everything Is Connected to Everything Else and What It Means.* Plume

2. Chakrabarti, S., Dom, B., Raghavan, P., Rajagopalan, S., Gibson, D., and Kleinberg, J. M. (1999) Mining the Link Structure of the World Wide Web. IEEE Computer, 32(8): 60-67

3. Chen, Z., Liu, S., Wenyin, L., Pu, G., and Ma, W. (2003) Building a Web Thesaurus from Web Link Structure. SIGIR

4. Chung, F. and Lu, L. (2003) The average distance in a random graph with given expected degrees. Internet Math. 1(1): 91-114

5. Emigh W, and Herring S. C. (2005) Collaborative Authoring on the Web: A Genre Analysis of Online Encyclopedias. In Proceedings of the 38th Annual Hawaii International Conference on System Sciences

6. Everything2 http://www.everything2.com/

7. Fixed Reference: Snapshots of WikiPedia http://july.fixedreference.org/en/20040724/wikipedia/Main_Page

8. Gross, J. L. and Yellen, J. (2003) Handbook of Graph Theory. CRC Press

9. Hanneman, R. A. and Riddle, M. Introduction to social network methods. http://faculty.ucr.edu/hanneman/nettext/

10. Hardman, L. and Edwards, D. M. (1993) Hypertext: Theory into Practice, chapter Lost in hyper-space: Cognitive mapping and navigation in a hypertext environment, pages 90105. Intel-lect. Oxford

11. Ishikawa, T. and Ishikawa, Y. (2005a) Structural Features of Knowledge Networks on the World Wide Web, in Proceedings of ANDI 2005

12. Ishikawa, T. and Ishikawa, Y. (2005b) Local Navigation in Knowledge Networks Based on Subsumption Relations of Webpage Sets, in Proceedings of WEIN 2005 (in Japanese)

13. Jain, A. K., Murty, M. N. and Flynn, P. J. (1999) Data Clustering: A Review, ACM Comp. Surv

14. Newman, M. E. J. and Girvan, M. (2004) Finding and evaluating community structure in networks. Phys. Rev. E 69, 026113

15. Masthoff, J. (2002) Automatic generation of a navigation structure for adaptive web-based instruction. In Proceedings of Workshop on Adaptive Systems for Web-Based Education, pp. 81-91

16. Palotai, Zs., Farkas, Cs. and Lrincz, A. (2005) Selection in scale-free small world. In Proceedings of. CEEMAS 2005 LNAI 3690: 579-582

17. Rosenfeld, L. and Morville, P. (1998) *Information Architecture for the World Wide Web*. O'Reilly

18. Stacy, A. W. (1995) Memory association and ambiguous cues in models of alcohol and marihuana use. Experimental and Clinical Psychopharmacology, 3, 183-194

19. Tangmunarunkit, H., Govindan, R., Jamin, S., Shenker, S., and Willinger, W. (2001) Network topologies, power laws and hierarchy. TR01-746, University of Southern California

20. Watts, D. (2003) *Small Worlds: The Dynamics of Networks between Order and Randomness* (Princeton Studies in Complexity). Princeton University Press

21. Wikipedia, the free encyclopedia http://en.wikipedia.org/

22. Writeup http://www.everything2.com/index.pl?node=writeup

23. Yahoo! http://www.yahoo.com/

# From Agents to Communities: A Meta-model for Community Computing in Multi-Agent System

Kyengwhan Jee[1] and Jung-Jin Yang[1]

School of Computer Science and Information Engineering, The Catholic University of Korea, Yeouido Post Office, P.O.Box 960, 35-1 Yeouido-dong, Yeongdeungpo-gu, Seoul, Korea Tel: +82-2-2164-4377, Fax: +82-2-2164-4777
sshine106@catholic.ac.kr jungjin@catholic.ac.kr

**Abstract.** The Multi-agent System, together with the ubiquitous computing technology, establishes an environment which provides services suitable to the given situation without the restraint of time and space. Naturally, a multi agent system should be enlarged and capable of accommodating communications between diverse agents, which requires the system design at an abstract level and the templetization of communication. Our work will present a meta-model of the community computing for the abstraction of agent system design, and a structure of community computing middleware leading to the extension of the system. The community computing meta-model suggests two types of community to perform both templetized and autonomous communications, and its transitive composition of communities makes the abstracted design of a large system possible. The middleware, presented along with the meta-model, serves as the framework for the ecological system of entities constituting the community computing.

## 1 Introduction

Along with the spread and rise in influence of embedded and mobile devices with limited computational power, there is an increasing demand for integrating the various different kinds of such devices in order to provide an environment where access to information and services is available in a seamless manner. Agent-based systems have a key role to play in the effort to provide and support such integration, since agents embody several of the required characteristics for effective and robust operation in ubiquitous environments [1][2]. The interaction of diverse agents is also necessary for the provision of adequate services in the ubiquitous computing environment. To accommodate this interaction, a multi-agent system should be enlarged and capable of having active communications between diverse agents and a formal methodology

K. Jee and J.-J. Yang: *From Agents to Communities: A Meta-model for Community Computing in Multi-Agent System*, Studies in Computational Intelligence (SCI) **56**,67–81 (2007)
www.springerlink.com                                   © Springer-Verlag Berlin Heidelberg 2007

at an abstract level is indispensable to effectively design and establish a large multi-agent system. Accordingly, a variety of researches have been made for the abstraction and reutilization of the agent design.

To analyze and design the agent system, Agent-Object-Relationship Modeling Language [3] was suggested. It offered an internal AOR model reexpressing the mental state of the agent and established an analytical model for the agent system by providing an external AOR model which consists of diagrams reflecting the static and dynamic views of the agent system. However, such a model does not suffice to design the system of group agents cooperating in a proactive fashion. In order for a multi-agent system to achieve a task at hand in a cooperative way, it is imperative that the agent system be capable of summoning agent groups with the ability to solve the assigned problems, for which the system should be also comprehensive of the latent interactions between agents and agent groups. Odell et al [4] proposed a meta-model to accommodate the requirements for given social interactions and to model agents, the role of agents, and agent groups based on the UML superstructure [5]. The meta-model enabled the agent system to summon agent groups executing its assigned task. And the group agents communicate with one another performing the designated roles. However, agents need to change their roles in the group autonomously under certain circumstances, and the meta-model posed does not deal with this matter in detail.

An agent is an individual with intelligence and autonomy, and these characteristics have been studied in many ways. Nikolic, in his research [6], has induced these characteristics from the perspectives of "how is a task performed in a given situation?" while the similar features have been derived by Ferber [7] from the animals with autonomous mental state. The given meta-model is similar to Nikolic's viewpoint, and the agents have to observe the exact orders and rules in performing tasks within their groups. And yet, our work, by including Ferber's viewpoint, provides a more general meta-model in which agents can autonomously change their roles in their groups. The agents following Nikolic's concept basically perform templetized communications, and such type, labeled Structured Community, only carries out communications suitable to their purpose of design. On the other hand, agents accepting Ferber's view, labeled Emergence Community, can perform communications extended to their intended purpose as well as autonomous ones.

In the next section the current research trends about Multi-Agent System and its related areas will be introduced, and the third section will present a meta-model for the design of an enlarged multi-agent system on the foundation of communities. Based on the presented meta-model, the fourth section will provide a middleware architecture that constitutes the framework for the ecological system of agents. The architecture will take advantage of utilizing the ontology repository [9] and lead to the prosperity of the community computing based on standardized MAS [8]. The fifth section will deal with the experiences of those experiments conducted in order to induce the requirements of meta-model and middleware structures.

## 2 Related Work

In order to construct the mega multi-agent it is necessary to improve the efficacy of agent design, embodiment and analysis by introducing the standardized language, manufacturing device and monitoring tool. In this section the previous research acquiring these elements is reviewed.

### 2.1 Agent-Oriented Software Engineering (AOSE) for massively multi-agent system

Up to now, there are many methodologies and models based on the organizational and social abstractions. A list of methodologies and their evaluations can be found in [10][11][12][13][14][15]. In this section, only two representatives of them, which are influential in AOSE literature, are discussed. Gaia is originally proposed by Wooldridge et al. [13] and extended by Zambonelli et al.[14]. Gaia is a complete methodology based on the organization metaphor for the development of MAS and views the process of analyzing and designing MAS as one of constructing computational organizations. The system can be open or closed, dynamic or static. The extensions are mainly based on three high-level organization abstractions and concepts such as organization rule to represent the global organization constraints, organization structure and organizational pattern that are helpful to specify and analyze the dynamic system and can be reused from system to system. The extensions of Gaia enriches the modeling capability in the all agents level with the three high-level organization abstractions and concepts in order to deal with the complex open MAS. However, it is still weak in the level of two or more agents acting in a coordinated way. AALAADIN [15] is an abstraction and generic MAS model and particularly focuses on the organization-centered MAS modeling in order to resolve the drawbacks of classical agent-centered technologies. ALAADIN assumes that systems to be developed are closed and have hierarchic structure that can be decomposed as a number of interrelated groups, and agents in the system are cooperative. It is obvious that AALAADIN has rich expression to model MAS of agents acting in a coordinated way. But it is weak in the single agent level and suitable for the closed system with cooperative agents, and especially with the hierarchic structure, for instance, enterprise information system.

## 3 A Meta-model for Community Computing

The ultimate aim of the meta-model is to actively organize circumstantial communities and provides the base structure for the performance of assigned tasks. Also, the presented model as in Fig. 1 is supposed to lead to the adaptive and circumstantial reconstitution of communities. The Role Classifier in the upper-left of Fig. 1 is an abstract classifier classifying roles of the community and the agent, and is specialized as Community Role Classifier and Agent Role Classifier. The Community Role Classifier classifies roles of the Community that are realized as the Structured Community and the Emergence Community. The community manager selects necessary community members on the basis of the role classifier, and allows its resources for constructing communities and retrieves them from the destroyed communities.

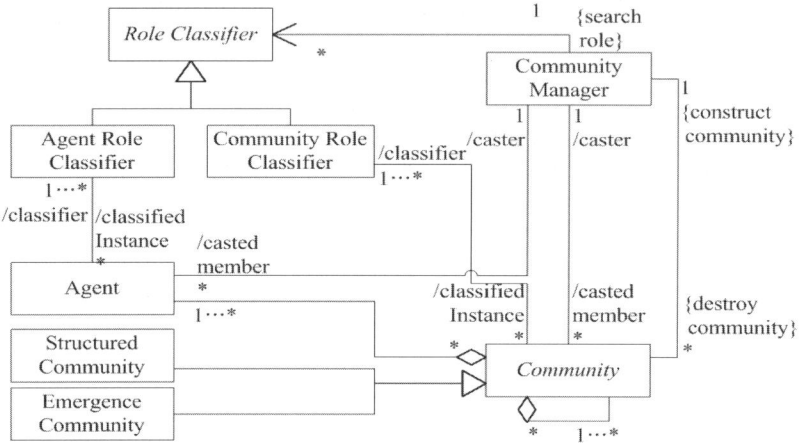

**Fig. 1.** A Meta-model for Community Computing

## 3.1 Agent and Agent Role Classifier

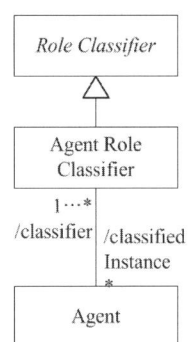

**Fig. 2.** Agent Role Classifier

Agent is the smallest unit constituting the community computing and cannot be divided further. Examples of such agents are the ones of communicating with a certain embedded system, handling passive entities or converting messages between agents. Should a task be executed in the community computing, it is necessary that communities be summoned under the control of the community manager[1].

The Agent Role Classifier in Fig. 2 is a classifier assorting agents, and agents are classified by at least one agent role classifier. The agent classifier can form the hierarchy of agents according to their features. For instance, agents handling physical devices are characterized by high cost and a long period of production/ change (e.g., mobile, embedded device handler). On the contrary, software agents for message conversion and protocol translation employed in communication can be produced at low cost and has a short period of production/

---

[1] This work does not deal with independently operating agents that do not belong to a community

change, which allows them to vary. These characteristics (e.g., cost, period of change) can be required in organizing communities, and the agent classifier can include corresponding information.

## 3.2 Community and Community Role Classifier

A community has a group communication of certain individuals, whose original purpose can be achieved by subdividing a certain task and constituting a community fitted to the subdivided purposes. A community can be cast with a community member to communicate with other members. As Fig. 3 shows, a community is classified by at least one community role classifier. Templetized community members communicate with one another in accordance with their assigned roles and can communicate with the outside of the community if it is appropriate to the role assigned.

## 3.3 Structured Community and Emergence Community

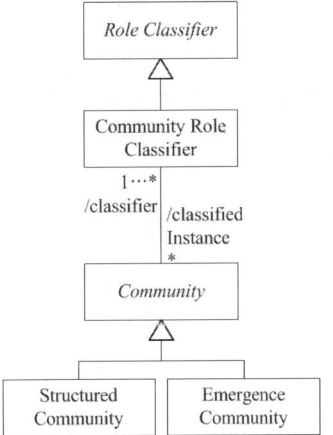

**Fig. 3.** Community and Community Role Classifier

A community, as shown in Fig. 3, is realized in two types. One is the Structure Community which carries out the communication templetized according to the designed purpose, and the other is the Emergence Community which communicates according to the dynamically assigned purpose of the community members. The Structured Community can be regarded as an agent group designed appropriately to the designer's intention. Likewise, the Emergence Community can be seen as a group that is permeated with the designer's intention, but includes arbitrary phenomenon caused by unpredictable situations and personality of community members. As shown in Fig. 3, a community is an Abstract and implemented as either the Emergence Community or the Structured Community.

The Structured Community is implemented on the purpose of multicasting those messages that have destinations among the community members. On the contrary, the Emergence Community is embodied on the purpose of broadcasting such messages, and it can understand, discard or misunderstand messages depending on the community member's capability. Hence, whereas the Structured Community can accomplish tasks in the order of the transfer of messages intended by the designer, the Emergence Community might transfer messages differently from the intention of the designer and cause an unpredictable condition following uncertain circumstances.

## 3.4 Community Manager

The smallest unit needed to complete a task in the community computing is Community. Agent is the smallest unit constituting the community computing whose main job is to contextualize the performance of behavior and the cognition of information, and to constitute a community to communicate with other agents or other communities.

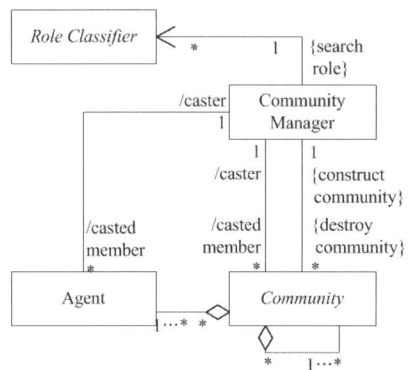

**Fig. 4.** Community Manager

The community manager administers resources consumed in the community computing and constitutes communities. As expressed in Fig. 4, the community manager casts communities or agents classified by the role classifier as the community members. Communities performing a task inform the community manager of their demise upon the accomplishment of their aims so as to return their resources.

To organize a community, the Community Description Model, which describes the community information, should be defined in formal language. The Community Description is drawn up on the basis of the Community Description Model and the drawnup message contains roles required in task performance and the information communicable with the context relevant to the task performance, and they become the ground of the casting. Besides, the community manager should be capable of reflecting the characteristics of the agents and communities, and execute scheduling in order to avoid a deadlock situation.

# 4 Community Computing Middleware Architecture

The Community Computing Middleware Library, in the lower-left of Fig. 5, is presented in the form of a library compatible with the FIPA-complaint MAS foundation. The Community Computing Middleware Library constitutes the framework for the community-oriented ecological system by deploying the Ontology Agent Factory which produces the ontology agent utilizing the ontology repository, the Wrapper Agent which transforms the existing objects and components into agents to perform the agent-oriented community computing, the conversion agent for the conversion of messages, and the Proxy Agent for the translation of protocols. The Community Observer is an agent that monitors communities and additionally functions to convert the Emergence Community into the Structured Community by monitoring the Emergence Community.

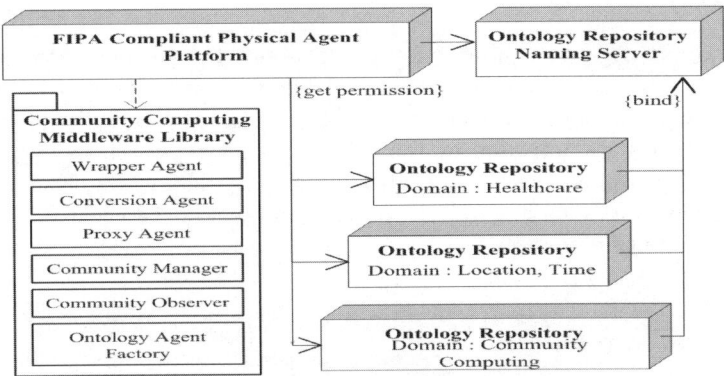

**Fig. 5.** Community Computing Framework Overview

## 4.1 Ontology Agent

The community manager casts agents or communities categorized as the role classifier, who are the members required for the organization of a community. The role classifier may have its own hierarchy depending on diverse strategies, and can be stored in its repository in various forms of contents according to its structure. The major information the community manager retrieves from the ontology repository is the community description, which is to be used for organizing a community, and the role classifier that is to be used for casting. The general methodology of the utilization of ontology can be obtained if the community manager circumscribes the type of necessary information to realize the community computing. To utilize the ontology repositories in diverse fields shown in Fig. 5, the ontology agent factory obtains the query and connection privilege optimized to the ontology structure by interacting with the Ontology Repository Naming Service.

For instance, the ontology agent factory can constitute an ontology agent by collecting the query, which enables the community manager to retrieve the community description and role classifier to be used, from the Community Computing Ontology Repository. Or it can do so by acquiring the privilege from the Location Ontology Repository or the Healthcare Ontology Repository in order to retrieve the context to be used in the community. The ontology agent produced subsequently is cast as a community member with the role of utilizing the ontology repository and communicates with other members. It is to ensure the access to a variety of ontology repositories that the ontology agent factory is used since the adaptability of the services offered by the community computing proposed in this paper is governed by the ontology structure.

## 4.2 Wrapper Agent

The Wrapper Agent acquires information about the required context, the related service shown in Yellow Page, and service description by the communication with the ontology agent. On the ground of the acquired information, passive entities are activated by the Object Broker expressed in Fig. 6. The Wrapper Agent plays the role of controlling diverse existing systems like those agents producing/spending agreed contexts and makes the agent-based community computing possible.

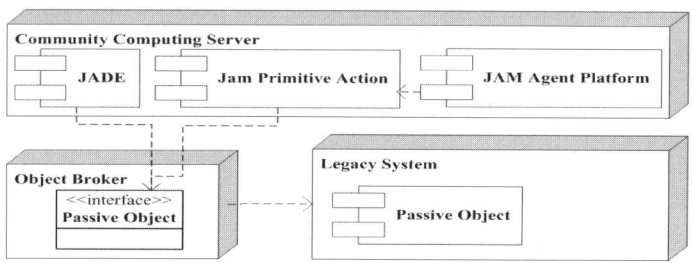

**Fig. 6.** Object Broker for Wrapper Agent

The ontology repository has to ensure the role of services provided by passive entities and the semantic interoperability of the produced/consumed data, as well as furthermore to function as a knowledge store for the performance of the Service Composition through the community computing.

## 4.3 Conversion and Proxy Agent

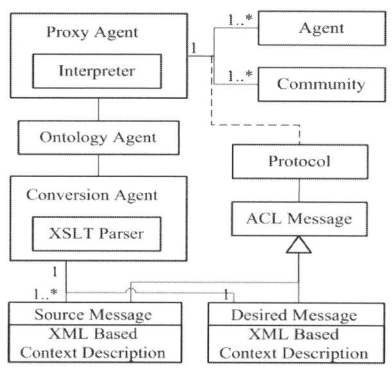

**Fig. 7.** Conversion Agent and Proxy Agent

A number of agents organizing a community produce and consume various types of context. However, an agent may not be provided with a context consumed to perform a certain function in diverse and constantly changing situations. On the other hand, it has to produce a context to be used to design an agent that shall play a new role. Likewise, the Conversion Agent, as shown in Fig. 7, functions to produce the necessary context out of the context produced and consumed by an agent performing a similar function, and it produces the intended context through such changing processes as filtration, merger, composition, translation, and calculation. As in the lower section of Fig. 7, a context is included in a message to be transferred and produces the object message out of at least more than one message. The conversion agent retrieves XSLT-oriented conversion rules, which are employed in the converting task, from the ontology repository and carries out conversion using the XSLT parser on the ground of the rules. The conversion agent is cast as a source-context-producing agent or as a community member together with the community.

The Proxy Agent, as depicted in the upper section of Fig. 7, intermediates translatable protocols between agents. The proxy agent retrieves translation rules described by the ontology agent and is cast into a community to enable the community members of different protocols to communicate with one another.

# 5 Experiment

An experiment with a virtual scenario was conducted to induce a meta-model of the community computing and the requirements of the architecture. The scenario is referred as [Scenario 1] in which communities are organized with the aim to keep a child safe.

## 5.1 Scenario for safety

[Scenario 1 - Safety] (Within the safe area) Mother Jiyeon Kim is currently busy in the kitchen. Her daughter Sooyun Park is merrily playing in the playground of the apartment complex with her friends. Mother was worried about her daughter playing outside alone so she had her daughter go out with a Smart Belt worn to ensure her safety.

(Level-1 dangerous area) Not long after, Sooyun goes out of the playground to play with her friends and moves to the Level-1 dangerous area. A community is organized to inform the mother at home of the Level-1 dangerous situation, and it notifies her through a kitchen appliance (an electric rice cooker) located close to Jiyeon's current location. Having seen the warning message, Jiyeon confirms that her daughter is within the apartment complex thinking that she does not have to worry and continues her work at the kitchen terminating the warned situation.

(Level-2 dangerous area) Sooyun moves to the Level-2 dangerous area from the Level-1 dangerous area. Another community is organized to inform Jiyeon of the Level-2 danger which does so by all means possible-e.g., a PDP TV in the living room, the Smart Kitchen Appliances (an electric rice cooker etc.). Accordingly, the Smart Kitchen Appliances send a text message along with a warning sound and the PDP TV is automatically turned on to display the current situation of Sooyun through the cameras installed at the Smart Streetlamps with its volume increasing gradually. Recognizing the situation, Jiyeon approaches the PDP TV only to confirm that Sooyun moves out of the apartment complex, which she judges to be a dangerous situation and reports. The community manager locates Hongim Choi and Sunghee Shin as available people around Sooyun among the off-line community members (a friendly group). Given the distance from Sooyun, the manager sends a message to the OMD[2] of Choi asking for help, who responds that he cannot help since he is busy with other matters. Then the manager contacts Sunghee Shin who willingly accepts to help. Upon the acceptance, the manager sends the current location and image of Sooyun to Shin's OMD and she brings Sooyun back home taking advantage of the information provided.

## 5.2 Sequence Diagrams of the Scenario in Multi-Agent System

Since not all the agents and devices can be implemented for the scenario, only the agents that will constitute communities and accomplish the end of the scenario are selected. Basing on the selected agents, Fig. 8 shows the sequence of the organization and the demise of the community when the child moved to the Level-1 dangerous area.

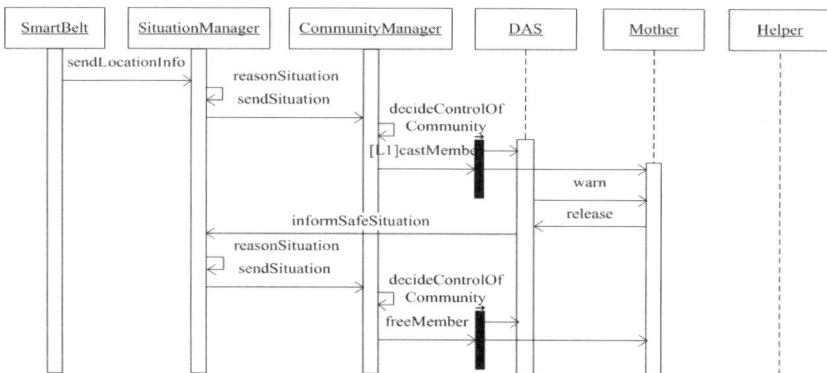

**Fig. 8.** Sequence Diagram for Level-1 Area

The Smart Belt Agent periodically transmits the local information of the child to the Situation Manager Agent. The Situation Manager Agent infers the current situation from the local information, the previous situation, and the community in activity provided. The Community Manager Agent receives the situational information from the Situation Manager Agent and constitutes a community organized to incite the safety of the child in the Level-1 dangerous area. Digital Appliance Server (DAS) raises an alarm through the kitchen appliances close to the location of the Mother agent, which, confirming that

---

[2] Organic Mobile Device. It is assumed as a set of personal and portable mobile devices in this paper

the child is within the apartment complex in the Level-1 dangerous area, informs the Situation Manager Agent that the situation is safe. The Situation Manager Agent infers the current situation from the received and local information, movement situation, and the community in activity and passes it on to the Community Manager Agent. Community Manager Agent can also extinguish the community in activity on the ground of the information given.

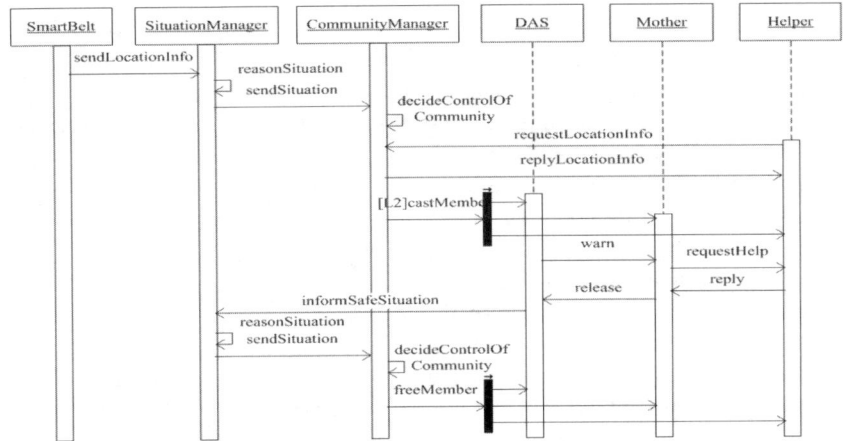

**Fig. 9.** Sequence Diagram for Level-2 Area

Fig. 9 presents the sequence of the community organization when the child moved to the Level-2 dangerous area. The Situation Manager Agent and the Community Manager Agent will do the same operation as in the Level-1 dangerous area. However, the community members and their activities are different. Unlike the former case, the community organized in the Level-2 dangerous area includes helpers near the child as well as all Wrapper agents of DAS (i.e. all home appliances) as its members. Moreover, more active alarm activities and diverse communications between the members incite the safety of the child. DAS warns the Mother agent of the child's danger by all home appliances available, and the Mother agent requests approximate agents to offer help. A helper takes the child to the Level-1 or a safe area, of which the Mother agent informs the Situation Manager Agent. Then the community is extinguished upon the activity.

### 5.3 Simulator API

Each agent is implemented from the two provided diagrams using a JAM [16] agent which is based on BDI Architecture. Jam agents, implemented in Java, employ behavior through the Primitive Actions defined, and they, as introduced in Fig. 6, call a remote object through the Object Broker. In this experiment, each agent is implemented as a client for the communication of agents and the creation and demise of communities, and it curbs simulators activated in the server.

The purpose of the simulator here is to create a simulated environment for the creation of communities fit to each situation, and to embody and monitor the communication and activity of the members. The simulator consists of passive objects activated by the Wrapper Agents. APIs of the passive objects in our simulator are described in Fig. 10 and implemented in Java. The creation and control of objects, monitoring message passing, and the creation/demise of communities are done through the simulator screen. The simulator provides translational message channels, and the method of message passing is conceptually the same as that of the Emergence Community broadcasting.

The embodied agents, like Ferber's view, performed autonomous communications as agents of independent mind, and there have been a lot of trials and errors in constituting their BDI. Although the number of agents is relatively small in the simulator, the value of Beliefs, which changes simultaneously in concept, and uncontrolled communications have sometimes shown entirely unintended results. Consequently, it has been learned that behaviors should be constituted in small units to encompass diverse uncertainties since the Emergence Community possesses the uncertainty each community member possesses at Behavior level. Additionally, it has also been found out that the Emergence Community can be reconstituted into the Structured Community by monitoring and tracking, and

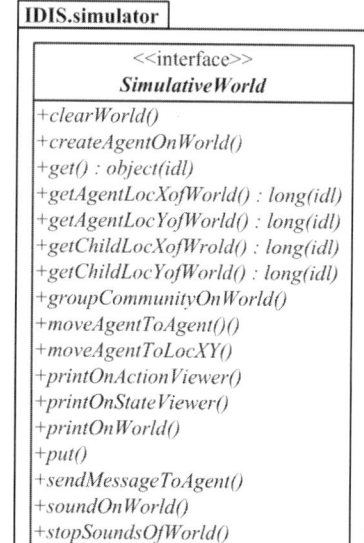

**Fig. 10.** Interface for Simulator

that the utilization of a tool that induces the Structured Community from the Emergence Community enables community computing more feasible.

## 5.4 Look and Feel for Simulator

Fig. 11 is the operating screen of the simulator and agents previously dealt with, of which the upper-left section shows the Control Panel that adjusts the position of the child, and the lower-left shows the Viewer that monitors the action of all agents. In the right side of the simulator is placed the Viewer that can observe the translational space of messages (i.e. Environment State), and at the center lies the screen. Each entity arranges its name and expression in pairs and identical community members are marked in the same colors.

In Fig. 11, the upper-right section is the house, the upper-left the Level-1 dangerous area, and the part below the road marks the Level-2 dangerous area.

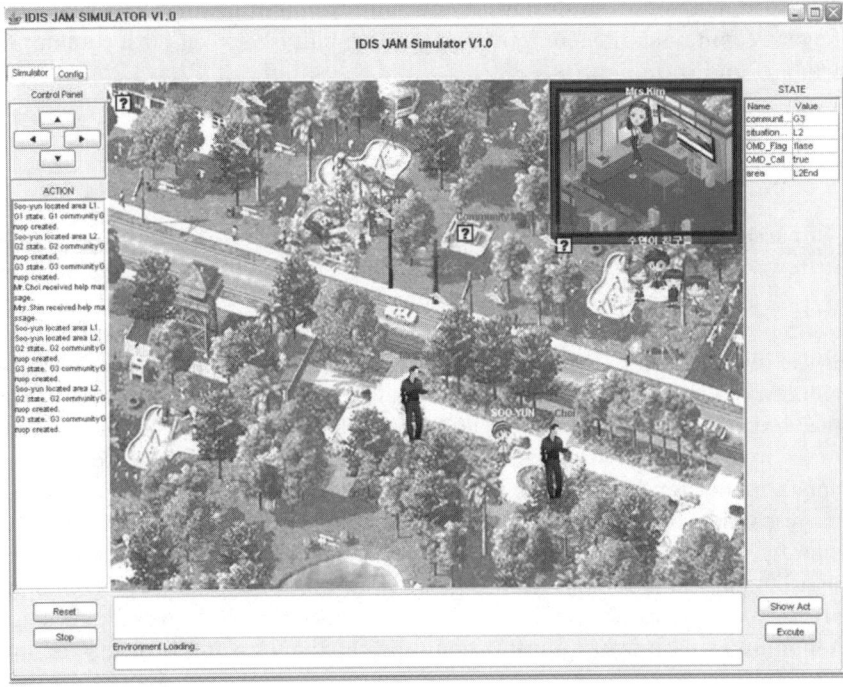

**Fig. 11.** Main Panel

The name of the child is written as "SOO_YUN" and the locations of her movements in different areas are observed through the Action Viewer. The entities in red in Fig. 11 and the OMD in the left side of Fig.12 are the members cast into the community when the child was in the Level-2 area. The panels

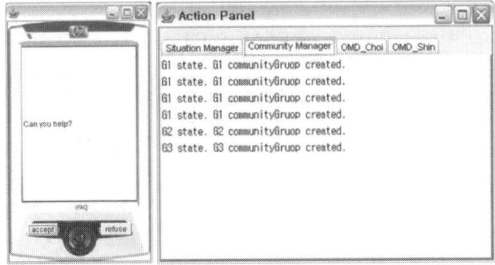

**Fig. 12.** OMD and Action Panel

in the right side of Fig. 12 are the Viewers for monitoring the actions of each agent and the left panel is the OMD. The OMD is cast into a community as a helper agent and was embodied to enable the user to choose to help or not by using the "accept" or "refuse" button.

Experiment of the simulator implemented enlightened helpful insights of community computing research. The control of passive objects through the Wrapper Agent and the context conversion through the conversion agent provide the basis of modeling a multi-agent system. Furthermore, the ontology repository, employed to deal with ambiguity of terminology and semantics of agent communications, improves interoperability between both community members and communities.

# 6 Conclusion

The paradigm of the software engineering is moving from the object-oriented one toward the agent-oriented one, and a variety of methodologies, attempting to solve a problem by employing agents, are being actively studied. The research on Multi Agent System, which can establish standard propositions for the interaction of agent systems, accept and operate these standards, also has been made actively. Our work, building on the preceding research in this area, induced a meta-model of the community computing and the requirements of the middleware architecture. While previous researches constituted a meta-model for the agent group reflecting the viewpoint of the designer, this research provided a meta-model susceptible to result in an uncertain outcome by guaranteeing the autonomy of the agent. The ultimate aim of the meta-model is to induce the change of knowledge through uncertain communications and support agent systems that can induce the evolution of knowledge. Communities with uncertainties should be capable of being transformed into communities free of those uncertainties, which will be the subject of the future research.

We have experimented general meta-models for performing given tasks through scenario setting, demo experiences and middleware architectures, and we have been able to perform the basic research constituting the middleware for the community computing. As its extension, our research currently runs on the Community Description Model for describing communities in a formalized form, and the Knowledge Description Model for describing a medium for knowledge in a formalized form. Each model, along with the protocol, should be capable of being accommodated into the ontology repository as contents for the community computing. Also, the Emergent Community can be used in research in a variety of fields and poses the necessity of monitoring tools and translators that converts them into the Organized Community.

*Acknowledgement.* This research is supported by Core Technology Development for Ultra Precision and Intelligent Machining to Generate Micro Features on Large Surfaces and by the department specialization fund, 2006 of the Catholic University of Korea.

# References

1. F. Bergenti and A. Poggi (2002) Ubiquitous Information Agents. International Journal of Cooperative Information Systems, 11(3-4), pp.231-244
2. T. Finin, A. Joshi, L. Kagal, O.V. Patsimor, S. Avancha, V. Korolev, H. Chen, F. Perich, and R.S. Cost (2002) Intelligent Agents for Mobile and Embedded Devices. International Journal of Cooperative Information Systems, 11(3-4), pp.205-230
3. G. Wagner (2003) The Agent-Object-Relationship Meta-Model: Towards a Unified View of State and Behavior, Information Systems, 28(5), pp.475-504
4. J. Odell, M. Nodine, and R. Levy (2005) A Metamodel for Agents, Roles, and Groups, Agent-Oriented Software Engineering (AOSE) V, James Odell, P. Giorgini, J?rg M?ller, eds., Lecture Notes on Computer Science volume (forthcoming), Springer, Berlin
5. OMG (2005) UML 2.0 Superstructure Specification, 05-07-04
6. I. Nikolic (2001) GeneScape: Agent Based Modelling of Gene Flow in Crop sand its implications for sustainability of Genetic Modification, MSc thesis
7. J. Ferber (1999) Multi-Agent Systems, Addison Wesley
8. FIPA, FIPA specifications, http://www.fipa.org
9. J. Pan, S. Cranefield, D. Carter (2003) A Lightweight Ontology Repository, Proceedings of the Second International Joint Conference on Autonomous Agents and Multi-Agent Systems (AAMAS 2003), ACM Press, ISBN: 1-58113-683-8
10. G. Weib (2003) Agent Orientation in Software Engineering, The Knowledge Engineering Review, 16(4), pp.349-373
11. O. Arazy and C. Woo (2002) Analysis and Design of Agent-Oriented Information Systems, The Knowledge Engineering Review, 17(3), pp.215-260
12. M. Kim et al. (1999) Agent-Oriented Software Modeling, In Proceeding of Sixth Asia Pacific Software Engineering Conference, pp.318-325
13. M. Wooldridge, N. Jennings, and D. Kinny (2000) The Gaia Methodology for Agent-Oriented Analysis and Design, International Journal of Autonomous Agents and Multi-Agent System, 3(3), pp.285-312
14. F. Zambonelli, N. Jennings, and M. Wooldridge (2003) Developing Multiagent Systems: The Gaia Methodology, ACM Transactions on Software Engineering Methodology, 12(3), pp.317-370
15. J. Ferber and O. Gutknecht (1998) A Meta-model for the Analysis and Design of Organizations in Multi-Agent Systems. In Proceedings of Third International Conference on Multi-Agent Systems, IEEE Computer Society, pp.128-135
16. Marcus J. Huber, Ph.D. (2001) Usage Manual for the Jam! Agent Architecture, http://www.marcush.net/IRS/Jam/Jam-man-01Nov01.doc
17. S. Johnson (2002) Emergence; The Connected Lives of Ants, Brains, Cities, and Software, New York: Scribner, ISBN:0684868768

# The effects of market structure on a heterogeneous evolving population of traders

Dan Ladley[1] and Seth Bullock[2]

[1] Leeds University Business School, University of Leeds, UK,
danl@comp.leeds.ac.uk
[2] School of Electronics and Computer Science, University of Southampton, UK,
sgb@ecs.soton.ac.uk

**Summary.** The majority of market theory is only concerned with centralised markets. In this paper, we consider a market that is distributed over a network, allowing us to characterise spatially (or temporally) separated markets. The effect of this modification on the behaviour of a market with a heterogeneous population of traders, under selection through a genetic algorithm, is examined. It is demonstrated that better-connected traders are able to make more profit than less connected traders and that this is due to a difference in the number of possible trading opportunities and not due to informational inequalities. A learning rule that had previously been demonstrated to profitably exploit network structure for a homogeneous population is shown to confer no advantage when selection is applied to a heterogeneous population of traders. It is also shown that better-connected traders adopt more aggressive market strategies in order to extract more surplus from the market.

## 1 Introduction

Understanding the centralised market has been one of the key aims of economic research for many years. Both the behaviour of the market and the traders within it have been intensely scrutinised in order to determine how they operate. Analytical studies, (e.g. [1]), experimental studies, (e.g. [2]) and empirical analysis (e.g. [3]) have all been employed in this attempt.

In addition to analytical, empirical and experimental results, the use of simulation and more recently multi-agent simulation[4] has become increasingly important [5, 6, 7, 8]. Multi-agent approaches have enabled the relationships between trader micro-behaviour and market phenomena to be modelled, which is often analytically intractable and experimentally time consuming. In virtually all of these micro studies, the market is assumed to occupy a single location. All bids and offers are submitted in the same place, where all others may see and respond to them. Not all markets, however, are like this. Retail markets, for instance, are spatially embedded and consequently impose

D. Ladley and S. Bullock: *The effects of market structure on a heterogeneous evolving population of traders*, Studies in Computational Intelligence (SCI) **56**, 83–97 (2007)
www.springerlink.com                                  © Springer-Verlag Berlin Heidelberg 2007

costs in terms of the time and effort that it takes to visit other traders and acquire information. As a consequence of this, it is usually impossible for a trader to visit *all* possible partners. Instead, the trader will probably restrict information gathering to nearest neighbours, or key operators in the market. In this case the market no longer has a central location to which information is submitted and, as a result, different traders within the market may have access to different histories of bids and offers.

It is not only spatially embedded markets that may limit the ubiquity of market information. Traders in a financial market have ready access to all trading information. However, in this case the shear quantity of information may segregate the market. The traders incur very little cost in gathering information, instead the main cost is that of analysis. Analysing information takes time, meaning that it may be impossible for a single trader to study and accurately respond to all of the information in the market in a fast enough manner. Traders are therefore likely to ignore some of the information available and fail to take it into account when making decisions. In effect the trader will not be hearing some of the information even though it is available in principle. One possible consequence of this is to focus the attention of traders on a small subset of market products, leading to specialisation.

There is, however, an important difference between these cases. Although a market may be segregated in terms of information flow, trade is not as restricted as it is in the spatially extended case. In either of these cases, however, assumptions about centralisation of market processes no longer hold. Different traders within the market have access to different histories of bids and shouts and, potentially, a propensity to deal with particular partners rather than others. These problems aren't necessarily limited to human traders. It is possible to conceive of markets that are sufficiently large, fast-moving, and complex that even computer programs would find it inefficient to analyse all information present, or consider trading with every agent in the market. Recently models have started to appear that examine these types of problems. For instance [9, 10] both examine trading scenarios that take place across networks, similarly [11, 12, 13], amongst others, consider the connected problem of trade network formation.

This paper aims to investigate the valuation of information within distributed markets. As has previously been described, traders in these markets will have access to different information sources and therefore different pictures of the market state. This will be particularly apparent if some traders are more connected than others, i.e., they have more information sources and/or trading partners. These better-connected traders are, on average, likely to have a better understanding of the market than those traders who are less well connected.

The effect of this imbalance is important because to some extent the degree to which a trader is connected can be altered by the trader itself. It is well known that resources must be expended to gather information and that properly analysing information takes time. In many situations it is possible for a trader to change the proportion of its resources dedicated to gathering and analysing information, however, it is important to know under which circumstances to do this.

In previous work [14] we have examined markets where both trade and information flow are restricted in a manner represented by an explicit, fixed network of possible agent-agent interactions. The network governed which agents were able to communicate with each other and, therefore, which agents were able to trade with each other. Importantly, this network was not complete (fully connected), i.e., some traders within the market could not communicate directly with others.

In this initial work we wished to gain an understanding of the value of information in a simple separated market so the market network was fixed. Traders were not permitted to change their connections during the simulation. In future we hope to develop this system so as to better understand the circumstances in which it is favourable to change connectivity. The market used for these simulations was very simple, it was not designed to reflect the intricacies of any particular distributed market. Instead it was designed to provide general insight into the valuation of information in separated markets. The results found could be applied to any markets where information cannot flow freely. This includes retail markets, OTC markets, and many others.

It was found that traders who were more heavily connected had a valuation that was significantly closer to the theoretical equilibrium price of the market. The better-connected a trader was, the more information sources it would have and so the more accurate an opinion it could generate. The quality of information a trader possessed was, therefore, directly related to its connectivity.[3] In order to exploit this knowledge a simple modification was made to each trading agent's learning rule so that they weighted information according to the difference between the sender's connectivity and their own. As a result a trader would place more weight on information it heard from traders more connected than itself, and less on information from traders less connected. It was shown that on average a population of traders gained an advantage from using this rule, and that this advantage was enjoyed mostly by the least connected individuals in the population. In fact, the most connected individuals suffered a slight drop in performance when adopting this rather crude fixed learning rule.

This experiment assumed a homogeneous population. All traders within the population had parameters drawn from the same distribution and so behaved in a very similar manner. It is not difficult to argue that traders with different connectivities might perform better by adopting different strategies in order to exploit their position within the market network. For instance, we

---

[3] Although in that case information quality was based on a traders connectivity there was no reason why it could not be decided by other factors in a real-world market, such as a company's reputation or size or the previous history of information received.

would expect that if the most highly connected individuals described above had had the choice, they would have chosen not to employ the learning rule that disadvantaged them, whereas those that were least connected may have chosen to use the rule more strongly in order to gain more benefit.

In order to allow such a heterogeneous population of traders it is necessary to individually specify each trader's parameters. One-way to do this is to hand tune every trader to find its optimal parameter set. However, the optimal parameter set for a particular trader is likely to depend on the parameter sets of the other traders within the market. In order to solve this problem it was decided to compete trading strategies against each other. A co-evolutionary genetic algorithm was designed to allow the evolution of competitive strategy sets.

## 2 Model

Trading networks were constructed in which nodes represented traders and edges represented bi-directional communication channels. There are many possible network configurations that could be investigated for their effect on market performance, including lattices, Erdős-Rényi random graphs, small worlds, and graphs resulting from preferential attachment. This paper will focus on the latter class of networks since they exhibit some interesting properties, including the presence of well-connected "hubs", that have an intuitive appeal in terms of real-world markets, where it would be expected that certain major shops or investment banks would be much better-connected than individual shoppers or investors in their respective markets.

We employ an existing preferential attachment scheme [15]. A network of $N$ unconnected nodes is gradually populated with $Nm$ bi-directional edges. In random order, each node is consulted, and allocated an edge linking it to a second node chosen according to probabilities calculated as $p_i = (n_i + \delta)^P$. Here, $P$ is the exponent of preferential attachment and remains constant, $n$ is the node's current degree (number of edges), and $\delta$ is a small constant (0.1 for all results reported here) that ensures unconnected nodes have a non-zero probability of gaining a neighbour. Self-connections and multiple connections between the same pair of nodes were not allowed. All probabilities, $p_i$, were updated after every edge was added. After $m$ cycles through the population, the network was complete. Note that every node will have a minimum of $m$ edges, and a maximum of $N - 1$.

Markets explored here have a relatively high preferential exponent of $P = 1.0$ in order to generate networks that display a wide range of degrees. For all results reported here, $m = 10$. Initial tests showed that if $m$ was significantly less than this value, the market failed to converge as few traders were able to trade with their limited number of neighbours i.e., it was separated.

The market mechanism operates in discrete time. Each time period, one active agent (one who is still able to trade) is selected at random to make an offer or a bid. The other agents in the market may only respond to that shout during that time period either to make a trade or ignore it. Once that time period has elapsed the shout is removed. Second, we limit an agent's ability to trade such that they are only able to make offers to, or accept bids from, their network neighbours. Each market was simulated for a fixed number of time steps.

## 2.1 Trading Agents

Here, the ZIP trading algorithm is used to govern trader behaviour. ZIP, or Zero Intelligence Plus, traders were created by Cliff and Bruten [7] in response to work by Gode and Sunder [5], who created the "Zero Intelligence" trading algorithm in some of the first agent-based market simulations. The Zero Intelligence algorithm was designed to be the simplest possible algorithm that would allow trade to occur in a market. Two types of Zero Intelligence trader were introduced. The first, unconstrained traders (ZI-U), choose shout prices at random from a uniform distribution across the whole range of possible prices permitted, disregarding any limit prices. It was found that markets populated by these traders exhibited none of the normal properties associated with markets, such as convergence to the equilibrium price. The second type of zero intelligence traders (ZI-C) were *constrained* in the range of prices that could be shouted. Shout prices were again drawn at random from a uniform distribution, however, this distribution was now constrained by a trader's limit price. In the case of sellers, shouts were constrained to be greater than the limit price, while in the case of buyers, shouts had to be less than the specified limit price. Importantly, markets populated by traders using this algorithm were shown to behave analogously to real markets in that they converged to the theoretical equilibrium price [5]. This was interpreted as indicating that the market mechanism itself was the most significant factor in market behaviour, and that the design of the trading algorithm was not as important. Cliff and Bruten [7], however, showed this to be incorrect, demonstrating that the convergence observed during each trading period was an artifact of the supply and demand schedules used by Gode and Sunder. They demonstrated that, for a certain type of supply and demand schedule that was close to symmetric, the probability distribution of likely ZI-C bids and offers would result in convergence to the mean price. They then performed simulations to verify these results with a broader range of supply and demand schedules. For nonsymmetric schedules, markets populated by ZI-C traders failed to converge, or converged to a non-market-equilibrium value.

The ZIP trader differs from the ZI-C trader in that it learns from the market. Each ZIP trader has a profit margin associated with its limit price. In the case of buyers, the profit margin is the amount by which they wish to undercut their limit price to make a trade, and in the case of sellers, it is the amount by which they wish to exceed their limit price. When a ZIP trader shouts, the price is constrained by its limit price and profit margin.

The trader uses the market's response to its activity (and the observable activity of others) to update its profit margin. For instance, buyers observe the bids made on the market and whether or not they are accepted and adjust their profit margin accordingly. The ZIP algorithm employs the Widrow-Hoff learning rule with momentum [16] to adapt these profit margins throughout each trader's lifetime, maximising for each trader the possibility of making a profitable trade (for full details of this algorithm, see [7]). This learning rule allows the traders to rapidly converge on the optimal price, while the momentum term allows blips in the market to be ignored. Unlike ZI-C, ZIP traders are capable of finding the market equilibrium under a wide range of supply and demand schedules.

Here each ZIP trader was initialised with a random profit margin drawn from a uniform distribution $[A_1, A_1 + A_2]$. Each trader was also initialised with a random learning rate drawn from a uniform distribution $[B_1, B_1 + B_2]$ and random momentum value drawn from a uniform distribution $[C_1, C_1 + C_2]$. The target price for the Widrow-Huff learning rule was calculated from $T = Fq + G$ where $q$ is the shouted price and $F$ is a value drawn from a uniform distribution $[1.0, 1.0 + D_R]$ for buyers and $[1.0 - D_R, 1.0]$ for sellers. $G$ is a random variable drawn from a uniform distribution $[0.0, D_A]$ for buyers and $[-D_A, 0.0]$ for sellers. Thus, $F$ provides a small relative perturbation to the target price and $G$ provides a small absolute perturbation.

### 2.2 Topological Learning Rule

The standard ZIP learning rule makes no distinction between the information it receives from different individuals. It was demonstrated [14], however, that there is a relationship between trader connectivity and accuracy of valuation. The following modification to the standard ZIP Learning rule was devised to take this imbalance into account.

The Widrow-Hoff rule currently includes a fixed learning rate which influences how quickly a trader is able to learn. In order for the traders to take account of information quality, the learning rate was modified so that instead of being fixed, the value would be calculated for each piece of information received. This alteration results in ZIP traders placing more weight on information obtained from well-connected individuals than from less well-connected individuals.

The function $f(s, r)$ was defined, where $s$ and $r$ are the sender and recipient of a piece of information (a shout).

$$f(s, r) = \begin{cases} 0.3 + \dfrac{0.2 \log \frac{E(s)}{E(r)}}{\log(R_{max})} & : E(s) \geq E(r) \\[3ex] 0.3 - \dfrac{0.2 \log \frac{E(r)}{E(s)}}{\log(R_{max})} & : E(s) < E(r) \end{cases}$$

The function, $E$, gives the number of neighbours (degree) of a trader, and $R_{max}$ is the largest ratio of edges between two adjacent traders within the market. $M$ is the midpoint of the function and $Q$ controls the range of values that it can take. This enhanced learning rule weights information according to relative connectivity within the market, i.e., the ratio of the sender's connectivity to the recipient's connectivity determines the learning rate. When the sender is more highly connected than the receiver the information received is more likely to be accurate and so more adaptation occurs. When the receiver is more connected, the receiver's current picture of the market state is likely to be more accurate than the senders and so less adaptation occurs. The value is normalised by the maximum ratio present in the market to prevent unnatural learning rates. Connectivity ratios are log-scaled to ensure that learning rate adaptation is sensitive to the small differences in connectivity that characterise most sender-recipient pairs in a network generated by a preferential attachment process (where there will be only a few very well-connected individuals).

The Widrow-Hoff "delta" learning rule was modified by removing the learning rate and replacing it with the function $G(s, r)$:

$$G(s, r) = \alpha f(s, r) + (1 - \alpha)L$$

Where $L$ was the original Widrow-Hoff learning rate. This function allowed simple control of how much importance the trading strategy placed on the enhanced rule.

## 2.3 Genetic Algorithm

In this experiment it was desirable for different trading strategies to compete against each other in order to examine the selection of trading strategies. A co-evolutionary system was designed in order to do this. As was noted by Cliff [17] the behaviour of a ZIP trader is governed by eight real valued parameters that may be expressed as a vector V: $V = [A_1, A_2, B_1, B_2, C_1, C_2, D_R, D_A]$. In the case of enhanced ZIP traders it was necessary to introduce three new parameters that controlled the function of the topological learning rule, therefore, the vector used was W: $W = [A_1, A_2, B_1, B_2, C_1, C_2, D_R, D_A, M, Q, \alpha]$ These parameters were used to form a real valued genotype in which each parameter was bounded to lie between zero and one.

From previous results [14] we know that a traders connectivity affects its profitability within the market. As a consequence a traders connectivity may also affect its optimal strategy, therefore, it was desirable for traders with different connectivities to be able to evolve their own strategies independently. In order to do this it was necessary to maintain multiple populations of genotypes. One method of doing this would be to have one population of traders for each possible connectivity, i.e., a population for traders with 99 connections, a population for traders with 98 connections etc. There is a problem with this system, although there are many examples of traders with few connections in the networks there are relatively few examples of traders with large numbers of connections. It would have required a prohibitively large number of trials in each generation to evaluate each strategy's fitness accurately. In order to avoid this problem each market was broken up into a fixed number of groups, $G_N$ (20 in all experiments reported here), sorted by connectivity. The $N/G_N$ most connected individuals formed one group, the next $N/G_N$ most connected individuals formed the second group and so on. Previous results [14] showed that traders with similar connectivities tended to achieve similar results and so could possibly employ the same strategy. This justified the formation of a set number of populations that would each contribute the same number of traders to each experiment.

To populate the groups $G_n$ populations were formed each of size $S_p$ (in all experiments reported here $S_p = 25$), each population corresponded to a particular group. In each trial $N/G_N$ members of each population were chosen at random to form each group. The members of these groups were then added to the network in the appropriate places. Every member of each population participated in $T_n$ trials each generation (in all experiments reported here $T_n = 40$). Due to the randomness present in the market and allocation of limit prices it was necessary to assess the traders fitness multiple times in order to attain a meaningful estimation. Strategy fitness was the average profit extracted by a strategy over all trials in that generation. The average profitability has an intuitive appeal, as in the real-world profitable strategies are more likely to survive and be copied. Standard roulette wheel selection with fitness proportionate weighting was used to select individuals for entry into the next generation. Mutation occurred at every locus of a selected genotype with probability $P_m$ ($P_m = 0.05$ in all experiments presented here). Mutation consisted of a perturbation of the locus by a value drawn from a uniform distribution $(-0.05, 0.05)$. If the mutated value was greater than one or less than zero then the mutation was discarded and the original value used. In accordance with the method employed by [17], single point crossover was performed with probability $P_c$ ($P_c = 0.3$ in all experiments presented here).

## 3  Results

Evolution occurred over 1000 generations. Each market was populated by 100 ZIP traders. Each trader was randomly allocated a limit price in the range $[1.00, 2.00]$, and either the ability to buy one unit or sell one unit of an unnamed indivisible commodity. Each market simulation lasted for 400 time steps. Markets were constrained by networks, constructed as described above, with $P = 1.0$ and $m = 10$, and all markets operated through the market mechanism described above. At the start of the experiment genotypes were initialised with parameters randomly chosen from the uniform distribution$[0, 1]$.

Figure 1 shows the average fitness of individuals within five different populations averaged over 24 experimental runs. The left figure shows the result of evolving standard ZIP traders, the right figure shows the result of evolving the enhanced ZIP traders. In both cases, strong trading strategies are quickly found by all populations (within approximately the first 40 generations). From this point onwards, however, the fitness levels of the populations remain approximately constant.

It should be noted that although the absolute fitness does not change after the first 40 generations this might not mean that the strategies are not continuing to adapt. Fitness is measured by the average amount of profit a trader makes. In these markets, however, the amount of profit available is fixed (though there is some small variation depending on the random distribution of limit prices). In order for one population of traders to increase its fitness it is necessary for it to become more profitable relative to another population. This is difficult to do as other population are simultaneously attempting to adapt their strategies to do the same. As in many co-evolutionary settings the trading strategies are continually adapting against each other and cancelling out each other's advantages. So although the fitness's may appear constant, the strategies may still be moving and changing in the strategy space [18].

In both experiments, corresponding populations attain similar fitness's. This indicates that some populations may have inherent advantages within the market and that the traders are not able compensate for these differences. It appears that the more heavily connected populations (low numbered) are able to exploit their connectivity advantage and extract the same amount of profit from the market in both cases.

Figure 2 deals with the deviation of the traders valuations over the length of a market experiment. These results were obtained by performing 10,000 market experiments using the final populations from each of the twenty-four genetic experiments. At each time-step the deviation from the equilibrium price of each trader's valuation was measured (traders that had already traded were not included in this measure). In previous work it was demonstrated that traders who were more heavily connected had valuations that were closer to the equilibrium price than those who were weakly connected. In this case, however, all traders quickly converge to equally good approximations of the equilibrium price. The addition of the topological learning rule does not appear to have any effect on the ability of traders to identify the equilibrium price. The convergence occurs at the same speed, and to the same level, both with and without the topological learning rule. (Note, this measure will never converge to zero as some traders have limit prices beyond the equilibrium price which bounds their valuation away from it).

Figure 3 shows the average learning rate for each of the final populations. In the case of the standard ZIP traders the learning rate is inversely proportional to the connectivity (r value < 0.01), i.e., more connected individuals have a lower learning rate than the less connected individuals. In the case of the enhanced ZIP traders the learning rate appears to remain approximately constant across the populations. The enhanced ZIP traders have on average a higher learning rate than the standard ZIP traders.

Figure 3 shows the average initial profit margin for each of the populations. The range indicated on the graph is the range from which each trader's initial profit margin is drawn. This value is then adapted throughout the course of the experiment as shouts are heard. In both cases there is a positive correlation with connectivity (r value $< 0.01$). i.e., the more connected a trader the higher the initial profit margin. The value of the initial profit margin does not appear to depend on the use of the topological learning rule.

The average weighting factor for the topological learning rule for each of the populations remains approximately constant at 0.5 across all populations. No population exploits the rule more than any other. It should be noted that the midpoint ($M$) and range ($Q$) of the rule remain approximately constant over all populations at 0.35 and 0.45. In both sets of experiments the remaining parameters were approximately constant across populations. The momentum parameters $B_1 = 0.35$ and $B_2 = 0.70$ and the perturbation parameters $C_A = 0.35$ and $C_R = 0.3$.

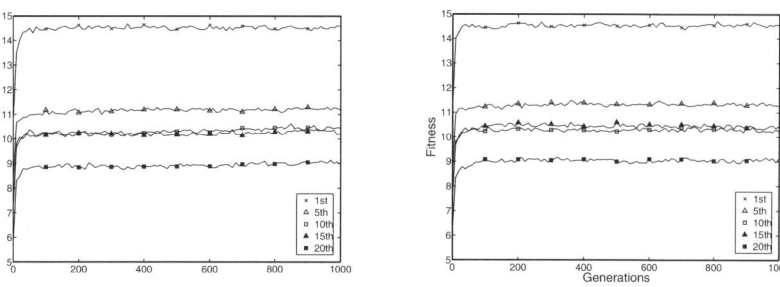

**Fig. 1.** Absolute fitness averaged over twenty four experiments for five of twenty populations ranked in decreasing order of connectivity for (left) standard ZIP traders, and (right) enhanced ZIP traders using a learning rule adapted to exploit market topology information

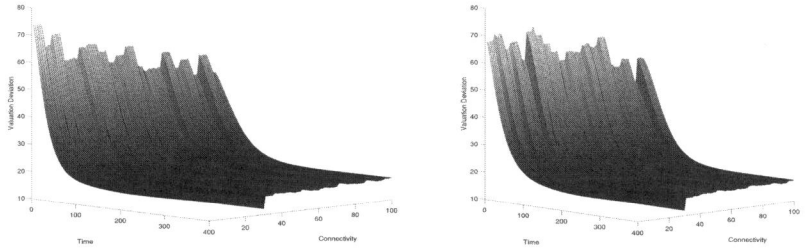

**Fig. 2.** Absolute deviation from optimum price averaged over 10000 runs for the final populations of each of 24 experiments. Traders ranked in decreasing order of connectivity for (left) standard ZIP traders, and (right) enhanced ZIP traders using a learning rule adapted to exploit market topology information

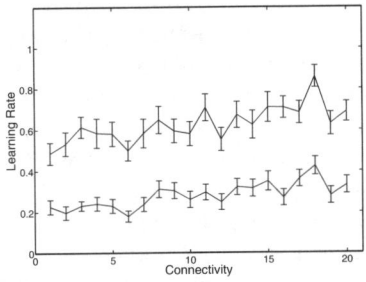
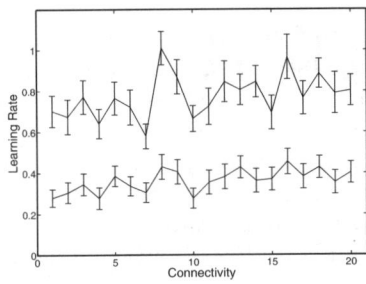

**Fig. 3.** Learning Rate for all members of the final populations of twenty four experiments, sorted in decreasing order of connectivity. (Left) Standard ZIP traders, and (right) enhanced ZIP traders using a learning rule adapted to exploit market topology information

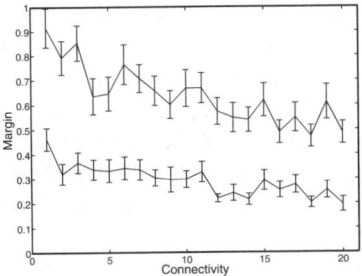
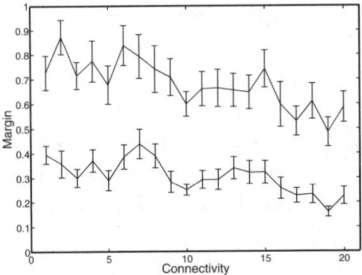

**Fig. 4.** Initial profit margin for all members of the final populations of twenty four experiments, sorted in decreasing order of connectivity. (Left) Standard ZIP traders, and (right) enhanced ZIP traders using a learning rule adapted to exploit market topology information

## 4 Discussion

This paper aimed to investigating effects of diverse trading strategies on trading behaviour in a structured market. Previous work had examined markets with a homogeneous population of traders [14]. In this paper this limitation was removed in order to allow a more realistic representation of a market where individual traders may develop their own strategies based on their circumstances and environment.

The first finding of this paper was that, traders who are better-connected on average start with higher initial profit margins. By having a higher initial profit margin the more connected traders are adopting more *aggressive* market strategies. A higher profit margin means the well-connected traders demand more from their trading partners and as a result will probably take a larger cut of the profit from the trade. They are effectively able to charge their trading partner a premium for the right to trade with them. How are they able to do this? There is no advantage for the partner in trading with the well-connected individual, the unit of goods bought or sold has the same value and no reputation is gained through trading with the well-connected individuals.

It is the market position that is exploited by the well-connected traders in order to increase their profits. A better-connected individual has many potential partners but it will only trade with one of them. Once it has traded all other traders are left with one less potential partner. If a trader is quick and agrees to the disadvantageous terms it is able to reliably make a trade and extract some (small) profit. If, however, it does not then it takes a chance on finding another partner who is more generous, or not finding any partner and so making no profit. The extent to which a well-connected individual may do this is governed by its connectivity, the better-connected an individual, the more likely it is that another trader will take the unreasonable terms and trade. Therefore, the better-connected an individual is the higher it can set its initial price and still expect to make a trade.

This paper demonstrated that the way in which traders learn is affected by their connectivity. In the case of standard ZIP traders the results (3, left), clearly show that learning rate is inversely proportional to connectivity. The less well connected a trader is the more it learns from each piece of information. This seems to be intuitively correct, if a trader receives a large amount of information it is possible to average over them all and place less weight on any individual piece. If information is relatively sparse then the trader must place more importance on each piece heard.

Figure 3 (right) shows the learning rates of the enhanced ZIP traders. It would appear from this graph that the enhanced ZIP traders do not follow the same trend, as there is virtually no slope present. When the effect of the topological learning rule is included, however, a slope appears. The weighting of the learning rule, the midpoint and range of the function remain constant across all population. The effect of this rule, however, is not the same for all populations. Traders in the more connected populations are more likely to hear information from a traders who is less connected than they are, i.e., those from a lower numbered population, and vice versa for those in less connected populations. This effect becomes larger towards extremes. When the effect of the topological learning rule is added to the fixed learning rate, a similar pattern is observed to that seen in the standard ZIP traders. Although the learning rule does not improve the trader's performance, it is used as an easy way to correctly shape the learning function.

The results also show that, in a market populated by standard ZIP traders, those traders who possess more connections are able to make more profit than those who have less connections, as demonstrated by their higher fitness (figure 1 left). The fitness remains almost unchanged with the addition of the topological learning rule (figure 1, right).

This is a very surprising result. Previous experiments [14] had suggested that the addition of the topological learning rule allowed less well connected traders to reduce the informational advantage of better-connected traders. As a result the performances gap between the best and least well-connected individuals could be narrowed. Accordingly it was expected that with the addition of the topological learning rule the less well-connected populations would gain a higher fitness and the better-connected populations a lower fitness than before. In appears, however, that this was not the case. The topological learning rule had no effect on the ability of traders to extract surplus from the market.

Figure 2 shows parameter sets are evolved such that after a small number of time steps, all traders, on average, have a valuation that is equally close to the equilibrium price. In previous experiments, using markets populated by homogeneous traders, those with more connections had a valuation significantly closer to the equilibrium price than those with fewer connections. The enhanced ZIP traders produce an almost identical result. They converge to a similar level in a similar amount of time.

The significance of this result should not be understated. It demonstrates that a very simple trading strategy, ZIP, is able to evolve to perform well in a structured market with limited information. Previously, in a homogeneous population, it was demonstrated that simple traders could use information quality in order to increase the accuracy of their valuations. In this case, a heterogeneous adaptive population was able to adapt a few simple parameters such that the deviation from the equilibrium price of the shouts was approximately equal across all individuals (figure 2). As a result information inequalities were no longer visible in the shouts and so could no longer be exploited. By adapting their parameters the ZIP traders were able to remove the effect of information inequalities from the market, all that remained was the effect of the market structure itself.

This fact has important consequences. If the structure of the market (the number of possible partners a trader has) is the only remaining factor that is unequal between traders then it must be this that causes inequalities in profits. Since this factor cannot be affected by the trading strategy then the design of more sophisticated trading algorithms may not be able to mitigate this effect. In this paper it was shown that the addition of an topological learning rule, that was previously demonstrated to be effective in structured markets such as this, had no effect on the fitness or valuation deviations. In other words the standard ZIP strategy was sufficiently advanced that it could find the competitive market equilibrium in these separated markets and that the addition of a more complex strategy could not improve on the result found for any group.

This leads to two possible hypotheses. First, in this simple trading scenario more sophisticated trading algorithms may not be able to significantly outperform the standard ZIP algorithm. Second, that no population of competitive traders (those that do not voluntarily give up possible profit) will be able significantly improve the performance of one population relative to another. In order to test these hypotheses more experiments must be performed. In particular experiments that pit the standard ZIP algorithm and the enhanced algorithm against each other and against other trading systems in the same market.

## 5 Conclusion

This paper has demonstrated that it is possible to evolve simple trading strategies to function well in structured markets. It has demonstrated that by tuning a few simple parameters it is possible for the traders to quickly remove any informational imbalances present within the market. Any differences in profits that remain are then solely due to differences in the number of possible trading partners that each trader has. It was also demonstrated that the addition of a more complex trading strategy that had proved to be effective in structured markets populated by homogeneous traders, had no effect on the distribution of profit within the market. The more advanced trading strategy was not able to mitigate the imbalances due to the market structure. The inherent imbalances were shown to be exploited by the better-connected traders, allowing more aggressive trading strategies to be employed successfully. This was primarily due to a larger number of potential trading partners.

These results also have important consequences for real markets. The fact that simple trading strategies may be tuned to remove informational imbalances indicates that this may also be true in real markets. The markets used in this experiment were simple, however, they do capture important features of real commodity markets i.e., there are shouts and trades that specify prices for goods of a known quality and volume. Some real markets, such as financial markets, are not entirely dissimilar from this, though it is accepted that real markets posses more commodities and more information sources. This work suggests that in heterogeneous populations of self-interested adaptive traders, such as those found in the real-world, it is possible to ignore informational advantages as a result of market connectivity. After a short amount of time, barring the effect of private knowledge, all traders should have an equally good valuation of the commodity. Given that maintaining trading connections probably has some cost, what then is the advantage of having multiple connections? The advantage comes from having more possible trading partners. The more trading partners a trader has the more aggressive it can be in its trading strategy and as a result the more profit it can make. So even though all traders may know the fundamental value of a commodity the structure of the trade network itself allows some traders to extract a higher price for that good than should theoretically be possible. In other words traders can exploit their market position to extract more surplus from a market than theory suggests they should.

# References

1. DeLong, J.B., Shleifer, A., Summers, L.H., Waldmann, R.J.: Noise trader risk in financial markets. Journal of Political Economy **98** (1990) 703–738
2. Smith, V.L.: An experimental study of competitive market behaviour. Journal of Political Economy **70** (1962) 111–137
3. Bouchaud, J.P., Mezard, M., Potters, M.: Statistical properties of stock order books: empirical results and models. Quantitative Finance **2** (2002) 251–256
4. Axtell, R.: Economics as distributed computation. In Deguchi, H., Takadama, K., Terano, T., eds.: Meeting the Challenge of Soical Problems via Agent-Based Simulation, Springer-Verlag (2003)
5. Gode, D.K., Sunder, S.: Allocative efficiency of markets with zero-intelligence traders: Market as a partial substitute for individual rationality. Journal of Political Economy **101** (1993) 119–37
6. Arthur, W.B., Holland, J.H., LeBaron, B., Palmer, R., Tayler, P.: Asset pricing under endogenous expectations in an artificial stock market. In Arthur, W.B., Durlauf, S.N., Lane, D.A., eds.: The economy as a complex evolving system II. Addison-Wesley (1997) 15–44
7. Cliff, D., Bruten, J.: Minimal-intelligence agents for bargaining behaviors in market-based environments. Technical Report HPL-97-91, HP Labs (1997)
8. Farmer, J.D., Patelli, P., Zovko, I.I.: The predictive power of zero intelligence in financial markets. Proceedings of the National Academy of Science **102** (2005) 2254–2259
9. Wilhite, A.: Bilateral trade and 'small-world' networks. Computational Economics **18** (2001) 49–64
10. Bell, A.M.: Bilateral trading on a network: a simulation study. In: Working Notes: Artificial Societies and Computational Markets. (1998)
11. Kirman, A.P., Vriend, N.J.: Evolving market structure: An ace model of price dispersion and loyalty. Journal of Economic Dynamics and Control **25** (2001) 459–502
12. Howitt, P., Clower, R.: The emergence of economic organization. Journal of Economic Behavior & Organization **41** (2000) 55–84
13. Pingle, M., Tesfatsion, L.S.: Evolution of worker-employer networks and behaviors under alternative unemployment benefits: An agent-based computational study. Technical Report 12327, Iowa State University (2005)
14. Ladley, D., Bullock, S.: Who to listen to: Exploiting information quality in a zip-agent market. In: To appear. (2005)
15. Noble, J., Davy, S., Franks, D.W.: Effects of the topology of social networks on information transmission. In Schaal, S., Ijspeert, A.J., Billard, A., Vijayakumar, S., Hallam, J., Meyer, J.A., eds.: Eighth International Conference on Simulation of Adaptive Behavior, MIT Press, Cambridge, MA (2004) 395–404
16. Widrow, B., Hoff, M.E.: Adaptive switching circuits. IRE WESCON Convention record **4** (1960) 96–104
17. Cliff, D.: Evolution of market mechanism through a continuous space of auction types. In: Computational Intelligence in Financial Engineering. (2002)
18. Cliff, D., Miller, G.F.: Tracking the red queen: Measurements of adaptive progress in co-evolutionary simulations. In: Proceedings of the Third European Conference on Advances in Artificial Life, London, UK, Springer-Verlag (1995) 200–218

# Analysis on Transport Networks of Railway, Subway and Waterbus in Japan

Takahiro Majima[1], Mitujiro Katuhara[1] and Keiki Takadama[2]

[1] Center for Logistics Research.
National Maritime Research Institute, Japan
{majy, kat}@nmri.go.jp

[2] Faculty of Electro-Communications
The University of Electro-Communications, Japan
keiki@hc.uec.ac.jp

**Summary.** Characteristics of network in the real world have attracted a number of scientists and engineers. Various findings are given from recent studies on the real world network, sometimes called complex network. The properties on the complex network are revealed mainly by early studies focused on the un-weighted relational network, in which there is no weight on links or vertices, such as distance or traffic amount. Meanwhile, studies investigating weighted network has begun to appear in recent years. In these papers the weight denotes distance between vertices. The knowledge of complex network seems to provide us useful information to design and construct the transport networks. Our final purpose is to find an algorithm providing effective and optimized waterbus and bus network expected to reduce traffic congestions and increase redundancy of transport system under disaster circumstances. Toward this goal, this paper starts by investigating five transport networks, one railway, three subways and one hypothetical waterbus lines in Japan and their combinations from the viewpoint of complex network. Furthermore the role of waterbus network is made clear using measures in terms of complex network.

## 1 Introduction

Characteristics of network in the real world have attracted a number of scientists and engineers [1-12]. Various findings are given from recent studies on the real world network, sometimes called complex network. The complex networks, such as *small world network* [1] or *scale free network* [4], lie between two extremes, i.e., *random network* and *regularly connected network* like lattice. Many complex networks in real world have interesting characteristics found in both random and regular networks simultaneously.

The properties on the complex network are revealed mainly by early studies focused on the *un-weighted relational network* [1-4], in which there is no weight on links or vertices, such as distance or traffic amount. Indeed, the original work of small world network [1-2] deals with World Wide Web (WWW), electrical power grid, relationship among film actors, and neural network of a worm as un-weighted network. Meanwhile, studies investigating *weighted* network has begun to appear in recent years [5-9]. In these papers the weight denotes distance brtween vertices. Furthermore, several studies [10-12] reported that a bipartite graph

T. Majima et al.: *Analysis on Transport Networks of Railway, Subway and Waterbus in Japan*, Studies in Computational Intelligence (SCI) **56**, 99–113 (2007)
www.springerlink.com

comprised of station and line groups, transformed from railway network in Euclidean space, has the nature of small world network. Some of them reported that the degree distribution of railway network in Euclidean space represents power law distribution [12].

The knowledge of complex network seems to provide us useful information to design and construct the transport networks. Our final purpose is to find an algorithm providing effective and optimized waterbus and bus network expected to reduce traffic congestions and increase redundancy of transport system under disaster circumstances. Toward this goal, this paper starts by investigating five transport networks, one railway, three subways and one hypothetical waterbus lines in Japan and their combinations from the viewpoint of complex network. Furthermore the role of waterbus network, that doesn't exist at this time but is expected to reduce traffic congestions, is made clear using measures in terms of complex network.

This paper is organized as follows. Firstly, measures corresponding to the types of network are defined in Section 2. Five transport networks are introduced in Section 3. Using the measures, the transport networks are analyzed in terms of complex network in Section 4. Finally, our conclusions are given in Section 5.

## 2   Network and Measures

In this section, two kinds of network are employed. One is the transport network in Euclidean space, where the vertex represents a station and the link represents a direct transport service between two stations to which the link connects. The other network is a unipartite graph as un-weighted network. The transport network involving lines in which trains or buses are plying can be seen as to have a unipartite structure. In addition to the network type, measures characterizing the network are summarized below.

### 2.1   Types of Network

According to Sienkiewicz's work [12], the network in Euclidean space is named to Space L as weighted network. On the other hand, the unipartite network is named to Space P and is dealt as un-weighted network. On one hand, the network in Space L provides information on efficiency in terms of the geographical distance between stations. On the other hand, the network in Space P provides the information in terms of line in which buses or trains ply.

#### 2.1.1   Network in Space L

The studies on the complex network typically have been dealing with a relational network owing to the convenience of mathematical treatments. Weighted networks like transport networks are analyzed recently. The network in Fig. 1(a) shows a sample of transport network in Euclidean space with train or bus lines. In the network, vertices, links, and weight represent the stations, routes of trains or buses in the Euclidean space and geographical distance respectively in this paper.

### 2.1.2  Network in Space P

The links in the un-weighted network just indicate whether or not the connection between two vertices exists. Considering lines, in which trains or buses ply, the network of Fig. 1(b) is transformed from Fig. 1(a) by generating complete graph among stations belonging each line. Complete graph is a graph that any vertex has direct links to the other all vertices. Thus, supposing a complete graph with vertices number of $N$, the degree of all vertices becomes $N$-1 and total number of links is $N (N -1)/2$. Degree is a number of links that one vertex has. In this paper, results of networks in Space P are obtained from the type of the network in Fig. 1(b).

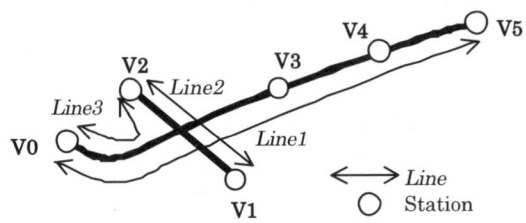

(a) Transport network in Euclidean space as weighted network in the Space L

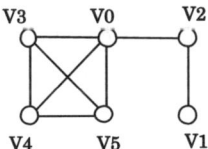

(b) Relational network in terms of lines as un-weighted network in the Space P

**Fig. 1.** Schematic drawings of sample of transport network

## 2.2  Measures

### 2.2.1  Efficiency

Latora proposed the measure called "Efficiency" [8-9] to consider the physical distance for weighted networks and to avoid a difficulty that the clustering coefficient (described in section 2.2.2) meets at terminal stations of transport networks.

Latora analyzed a transport network of subway in Boston with the efficiency. The efficiency has both behaviors like characteristic path length and clustering coefficient of the small world network described in the next section. The efficiency is defined as the following equations.

$$E(\mathbf{G}) = \sum_{i \neq j \in \mathbf{G}} \varepsilon_{ij} \qquad (2.1)$$

$$\varepsilon_{ij} = \frac{1}{d_{ij}} \qquad (2.2)$$

In these equations, $\mathbf{G}$ is a network, $\varepsilon_{ij}$ is the efficiency that is inversely proportional to the shortest distance, $d_{ij}$ between vertex $i$ and $j$. In the case of un-weighted network, $d_{ij}$, is number of links along the shortest path between vertex $i$ and $j$ and it becomes a geographical distance along the shortest path between vertex $i$ and $j$ for weighted network. If there is no path to connect vertex $i$ and $j$, $d_{ij}$ is infinity. Global efficiency, $E_{glob}$, and local efficiency, $E_{loc}$, are obtained from Eqs. (2.3), (2.4) respectively with above $E$.

$$E_{glob} = E(\Gamma)/E(\Gamma_{id}) \qquad (2.3)$$

$$E_{loc} = \frac{1}{N} \sum_{i \in \Gamma} \{ E(\Gamma(v_i)) / E(\Gamma_{id}(v_i)) \} \qquad (2.4)$$

In these equations, $\Gamma$ and $\Gamma(v_i)$ denotes the whole network with $N$ vertices and a sub-graph composed of neighbor vertices of vertex $i$ respectively. Subscript $id$ means idealized network comprised of complete graph among all vertices belonging. Thus, $d_{ij}$ of any ideal network becomes 1 for un-weighted network and becomes direct distance in Euclidean space for weighted network. It is clear by definition that global or local efficiency ranges from 0 to 1. When direct link for all pairs of vertices exists, the efficiency for un-weighted network becomes 1 and the efficiency for weighted network also becomes 1 when all pairs of vertex are connected by the straight links in the Euclidean space.

### 2.2.2   Characteristic Path Length, Clustering Coefficient and Diameter

The small world network [1-2], which is one kind of complex networks, is recognized by two conditions; small characteristic path length and large clustering coefficient. The characteristic path length, $L$, is the number of links in a route connecting two vertices with minimum steps, and averaged over all pairs in the network. The diameter, $D$, is defined as not average but maximum steps in the network. The physical meaning of clustering coefficient, $C$, is a ratio of number of existing triangle that includes vertex $i$ to the possible number of triangle computed by $k_i(k_i-1)/2$,

1)/2, where, $k_i$ is degree of vertex $i$. Mathematical formula of clustering coefficient, $C$, is described below.

$$C = \langle C_i \rangle \qquad (2.5)$$

$$C_i = \sum_{m,n \in \Gamma(v_i)} a_{m,n} \Big/ k_i (k_i - 1) \qquad (2.6)$$

In these equations, $C_i$ is clustering coefficient of vertex $i$, Symbol $<>$ means taking average over all vertices in the network. $a_{mn}$ is the $m, n$ element of adjacency matrix. Adjacency matrix stands for direct relations between two stations. The $ij$ element $\{a_{ij}\}$ of adjacency matrix stands for the existence of the direct links between vertex $i$ and $j$. If it exists, elements $\{a_{ij}\}$ becomes 1 and 0 otherwise. $\Gamma(v_i)$ is a sub-graph composed of neighbor vertices of vertex $i$ as mentioned above.

### 2.2.3  Eigenvector Centrality

Eigenvector centrality is a measure of an importance of a vertex in a network in terms of topology. Suppose that vertices have something that propagates to neighbor vertices through the links and it is represented by a vector, $\mathbf{x}$. The $i$ th element of the vector corresponds to the $i$ th vertex. Making product of multiplying adjacency matrix representing the direct relation between two stations and the vector yields recursively the something itself with factor of inverse of $\lambda$. This definition becomes eigenvector equation as is shown in Eq. (2.7) and the "something" represented by the vector $\mathbf{x}$ is called as eigenvector centrality. $\lambda$ and $\mathbf{x}$ in Eq. (2.7) stand for the maximum eigenvalue and corresponding eigenvector of an adjacent matrix, $\mathbf{A}$, respectively. Applying the eigenvector centrality to the adjacent matrix in Space P, the rank of centrality of stations can be obtained. This ranking evaluates an importance of stations from the viewpoint of architecture of train or bus lines.

$$\mathbf{x} = \frac{1}{\lambda} \mathbf{A} \mathbf{x} \qquad (2.7)$$

## 3  Transport Networks

Five transport networks, one railway, three subways, and one hypothetical waterbus line, are analyzed in terms of complex network using the measures described in the previous section. The analysis is typically performed around Tokyo Metropolitan, because Tokyo is the most populated area in Japan which population is over 10 millions and suffers from traffic congestions and commuter rushes chronically. The location of stations and links are collected from the information on web sites [13-15].

## 3.1  Railway Network

Fig. 2 shows the railway network in the Euclidean space of East Japan Railway Company (Hereinafter described as JR) within and around Tokyo. Train lines of 24 running into and through Tokyo are selected for our investigation. Number of stations, $N$, is 371.

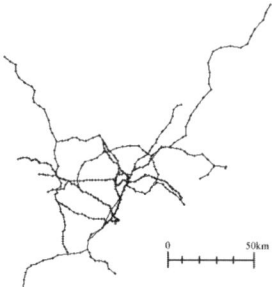

**Fig. 2.** Railway network of Japan Railway Company within and around Tokyo Metropolitan. Twenty four commuter lines running into or through Tokyo, Total number of stations is 371

## 3.2  Subway Networks

Fig. 3 shows subway network in Tokyo comprised of 12 train lines. The covering area is much smaller than that of JR lines. The network, however, encompasses the center of Tokyo. Total number of station, $N$, is 211. In addition to the sub-way network in Tokyo, those in Osaka and Nagoya, both of them are major cities in Japan, are analyzed simultaneously in this paper.

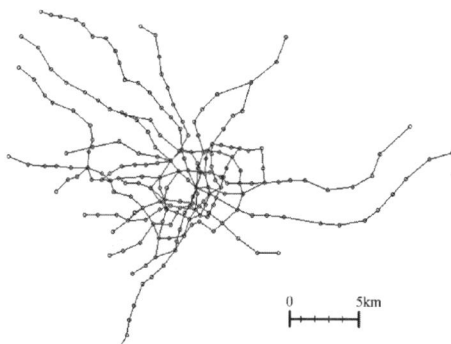

**Fig. 3.** Subway network in and around Tokyo Metropolitan comprised of 12 lines. Total number of station, $N$, is 211

## 3.3  Waterbus Network

The Tokyo metropolitan and around the region most populated area in Japan have been suffering from heavy traffic, congestions and commuter rushes on transport networks. The waterbus network, which doesn't exist at this time, is considered as hypothetical network in this paper. The waterbus line, however, may contribute to the more efficient transport network from the viewpoint of modal shift or reducing the commuter rushes. Several projects are under way [16], in which the combination of the river system and ships is utilized effectively as a substitute of road network suffered from chronic traffic congestion. Considering the fact that earthquakes occur frequently in Japan, road networks are collapsed when an earthquake causes massive destruction. The transport mode using rivers and ships is also being ex-pected to contribute to increasing a redundancy of the logistics system under disaster circumstances.

Fig. 4 shows network composed of Arakawa, Sumida and Onagi Rivers running through Tokyo. There are 24 possible waterbus stations in these rivers.

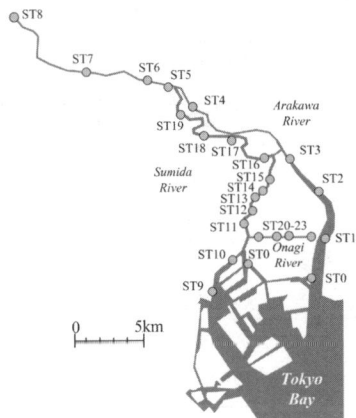

**Fig. 4.** Rivers in Tokyo, Arakawa, Sumida and Onagi Rivers. There are 24 possible water-bus stations in these rivers

## 4  Results and Discussions

### 4.1  Space L

The computed measures, global and local efficiency, $E_{glob}$ and $E_{loc}$ are summarized in Table 1 for Space L. In addition to five transport networks and one combined network of JR and subway in Tokyo, results for a subway in Boston obtained from literatures [9][11] is also described in the table. Combination network above means the transport networks of JR and subway in Tokyo combined at stations those are belonged to in common. There are 37 stations those JR railway and subway in Tokyo have in common. The weighted and un-weighted $E_{glob}$, weighted and un-weighted $E_{loc}$, and degree distribution are discussed below.

- $E_{glob}$ for Un-weighted Network

Table 1 shows that global efficiency for un-weighted network is small, approximately order of 0.1 except for waterbus. This value means that there are about 10 stations on average between two stations. Fig. 5 shows the relationship between the number of stations and $E_{glob}$ for un-weighted network. Un-weighted $E_{glob}$ decreases when $N$ increases. Supposing the network in Fig. 6 that is highly simplified image of transport network composed of outer circle and straight lines running through the center of the circle, all networks in this paper (except for Boston) have similar topology to Fig. 6. If the number of stations increases by inserting stations between existing ones, station with two degrees dominates the network and reduces the un-weighted $E_{glob}$. This is because the un-weighted $E_{glob}$ depends on the inverse number of stations between specified two stations. This explains the phenomenon indicated by Fig. 5 and relatively large value of waterbus network.

**Table 1.** Summary of computed measures for Space L

| Mode | Symbol | Station#<br>$N$ | Un-weighted | | Weighted | |
|------|--------|------|-----------|-----------|-----------|-----------|
| | | | $E_{glob}$ | $E_{loc}$ | $E_{glob}$ | $E_{loc}$ |
| Railway | JR | 371 | 0.075 | 0.026 | 0.78 | 0.033 |
| Subway | Tokyo | 211 | 0.14 | 0.019 | 0.70 | 0.024 |
| Subway | Osaka | 100 | 0.16 | 0 | 0.72 | 0 |
| Subway | Nagoya | 82 | 0.17 | 0 | 0.80 | 0 |
| Subway | Boston[9][11] | 124 | 0.10 | 0.006 | 0.63 | 0.030 |
| Combination | JR+Tokyo | 545 | 0.086 | 0.024 | 0.75 | 0.030 |
| Waterbus | WB | 24 | 0.34 | 0.25 | 0.77 | 0.20 |

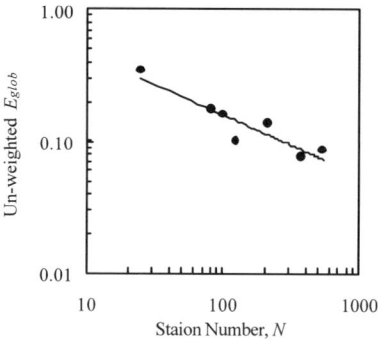

**Fig. 5.** Relationship between network size (total station number) and un-weighted global efficiency. All networks in Table 1 is plotted

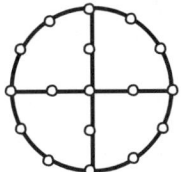

**Fig. 6.** Image of highly simplified transport network

- $E_{glob}$ for Weighted Network

The global efficiency for weighted network indicates large efficiency around 0.74. Furthermore, unlike un-weighted $E_{glob}$, the weighted $E_{glob}$ doesn't depend on the size of networks. It means that the transport network in real world is about 30% less efficient than the ideal network in which any pair of stations is connected by straight line in the Euclidean space. By definition, the global efficiency for weighted network has been already normalized with idealized network that has same network size. Thus, we can compare the global efficiency among networks that have different network size. The weighted $E_{glob}$ for the simplified network in Fig. 6 also doesn't depend on the network size strongly like observed values.

- $E_{loc}$ for Weighted and Un-weighted Network

The local efficiencies for both weighted and un-weighted network are extremely small except for waterbus. Especially local efficiencies for subway in Osaka and Nagoya are 0. This means by definition that triangle comprised of three stations does not exist at all. In other words, as mentioned in the Latora's paper [8], our transport networks result in effective in terms of global efficiency in the Euclidean space with sacrificing the fault tolerant. Thus, the same phenomenon to the precedent work is also recognized in the transport network in Japan.

- Degree Distribution

Fig. 7 shows cumulative degree distribution of subway network in Tokyo integrated from degree of $k$ to maximum of $k$. To smooth the fluctuation of probability distribution, $p(k)$, the cumulative distribution, $P(k)$ is employed. In these figure, $k = 1$ means terminal stations. Although the range of degree is limited to less than one decade, it seems that the distribution follows the power law distribution, $p(k) \propto k^{-\gamma}$, like scale free network as is shown in the Sienkiewicz's study [12]. Thus the same phenomenon is also observed in the transport network in Japan.

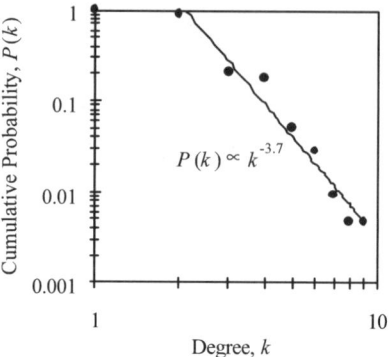

**Fig. 7.** Cumulative Degree Distribution of subway in Tokyo. The degree distributions seems to be power law distribution representing scale free network, although the range of degree is limited to less than one decade

## 4.2  Space P

The computed measures, clustering coefficient, $C$, characteristic path length, $L$, and diameter, $D$, in Space P is summarized in Table 2. Comparing the total number of stations, the diameter or characteristic path length of the network in Space P in Table 2 is extremely small. In the case of subway in Tokyo, the diameter is 3 and characteristic path length is 2. It denotes that we shouldn't change trains more than two times to reach any destination and averaged time for changing train is once within the network. Furthermore, the clustering coefficients become large value around 0.9. Thus, the networks in Space P satisfy the conditions that small world networks require and it matches with the results in the literature [12].

**Table 2.** Summary of computed measures for Space P

| Mode | Symbol | Station# $N$ | Line# $M$ | Unipartite Graph $C$ | $L$ | $D$ |
|---|---|---|---|---|---|---|
| Railway | JR | 371 | 24 | 0.92 | 2.6 | 6 |
| Subway | Tokyo | 211 | 12 | 0.89 | 2.0 | 3 |
| Subway | Osaka | 100 | 8 | 0.92 | 2.1 | 4 |
| Subway | Nagoya | 82 | 6 | 0.92 | 1.8 | 3 |
| Subway | Boston[9][11] | 124 | 8 | 0.93 | 1.8 | 3 |
| Combination | JR+Tokyo | 545 | 36 | 0.89 | 2.6 | 6 |
| Waterbus | WB | 24 | 8 | 0.89 | 2.0 | 4 |

Fig. 8 shows cumulative probability distribution in Space P. The left figure represents the degree distribution while the right represents centrality distribution. The degree distribution in Fig. 8 becomes exponential distribution, $p(x) \propto e^{-ax}$, it differs from that in Space L following power law like distribution. Not only the degree distribution but also the centrality distribution obeys the exponential distribution as shown in right figure in Fig. 8. It is known that the exponential distribution appears in random network. It implies that the stations in Space P involving consideration of lines are connected to another station randomly rather than the preferential attachment constructing scale free network [4]. Recalling the network in Fig. 6, the outer circle is connected to the ends of straight lines. Thus, it does not pay attention to the degree of stations.

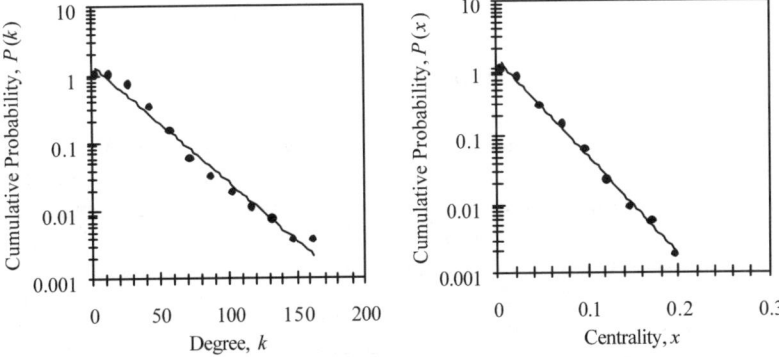

**Fig. 8.** Cumulative probability distribution of degree (left figure) and centrality (right figure) obtained from combined network of JR railway and subway in Tokyo

Table 3 shows the top 10 stations of ranking of centrality in the combined network of JR and subway in Tokyo. Although the positive correlation is observed between degree and centrality, it doesn't need that the station with the higher degree has the higher ranking in terms of centrality. Since the stations, such as Higashi-Nakano or Ryogoku, are not major stations, people living in and around Tokyo may feel peculiar about the ranking in Table 3. The reason is that stations with the high rank have lines connecting to a number of another lines. For instance, above two stations have both Sobu and Oedo lines whose number of transfer stations are the second and third highest among lines investigated in this paper. Thus, passengers at those stations can reach to most stations without changing train or with changing once.

**Table 3.** Ranking of stations in terms of centrality in Space P (combined network of JR and subway in Tokyo)

| Rank | Station | Degree | Clustering | Line # | Centrality |
|------|---------|--------|-----------|--------|-----------|
| 1 | Shinjuku | 164 | 0.216 | 7 | 0.200 |
| 2 | Tokyo | 167 | 0.196 | 8 | 0.171 |
| 3 | Akihabara | 120 | 0.289 | 5 | 0.161 |
| 4 | Yoyogi | 97 | 0.395 | 3 | 0.158 |
| 5 | Iidabashi | 126 | 0.266 | 5 | 0.148 |
| 6 | Shimbashi | 116 | 0.262 | 6 | 0.127 |
| 7 | Ueno | 126 | 0.267 | 6 | 0.120 |
| 8 | Higashi-Nakano | 71 | 0.555 | 2 | 0.115 |
| 9 | Ryogoku | 71 | 0.555 | 2 | 0.115 |
| 10 | Kanda | 85 | 0.371 | 4 | 0.115 |

## 4.3   Combination Network with Waterbus

The waterbus stations are made to be identical with railway or subway stations when the distance between them is less than 500m. It is assumed that all waterbus stations identical with stations in railway or subway network are connected only one waterbus line. Thus, it provides service for passengers to travel their destination without changing waterbus within the waterbus network. There are identical waterbus stations with stations of three for JR, six for subway in Tokyo and seven for JR, sub-way combined network.

Computed measures of Space L on all combinations with the waterbus network are summarized in Table 4. Table for measures of Space P is omitted because waterbus doesn't affect to them substantially.

**Table 4.** Summary of computed measures for combination network in Space L. The subway in the table is Tokyo and the symbol, "WB" is abbreviation of waterbus network

| Mode | Symbol | Station # $N$ | Un-weighted $E_{glob}$ | $E_{loc}$ | Weighted $E_{glob}$ | $E_{loc}$ |
|------|--------|---------------|------------------------|-----------|---------------------|-----------|
| Railway | JR | 371 | 0.075 | 0.026 | 0.78 | 0.033 |
| Subway | Subway | 211 | 0.14 | 0.019 | 0.70 | 0.024 |
| Combination | JR+Subway | 545 | 0.086 | 0.024 | 0.75 | 0.030 |
| Combination | WB+JR | 371 | 0.076 | 0.026 | 0.78 | 0.033 |
| Combination | WB+Subway | 211 | 0.14 | 0.024 | 0.70 | 0.031 |
| Combination | WB+JR+Subway | 545 | 0.086 | 0.027 | 0.75 | 0.033 |

The measures associated with $E_{glob}$ aren't affected by the combination with waterbus network. On the other hand, $E_{loc}$ for subway and combination network of JR and subway increase, even though the number of waterbus stations identical with stations in another network is very small. For example, the weighted $E_{loc}$ for subway network is 0.024 and it becomes 0.031 when waterbus network connects. However weighted $E_{glob}$ doesn't change.

Table 5 summarizes the result of centrality of stations to which the waterbus connects. Centralities of some stations improve by several dozen percent and the rank of the stations becomes higher drastically. The centrality of the Honjoazuma-bashi station of subway network increases by 45% and its rank improves from 175 to 86 owing to the waterbus network. Thus, the role of waterbus is to contribute not to $E_{glob}$, but to increasing the redundancy and centrality of the local network around stations to which the waterbus connects.

**Table 5.** Summary of centrality and rank of stations to which the waterbus network connects

| Station Name | JR | | JR + WB | | Centrality |
| | Centrality | Rank | Centrality | Rank | Increase |
| --- | --- | --- | --- | --- | --- |
| Etchujima | 0.035 | 158 | 0.038 | 156 | 7% |
| Ryogoku | 0.085 | 51 | 0.086 | 39 | 1% |
| Asakusa | 0.011 | 227 | 0.013 | 218 | 25% |

| Station Name | Subway | | Subway +WB | | Centrality |
| | Centrality | Rank | Centrality | Rank | Increase |
| --- | --- | --- | --- | --- | --- |
| Ryogoku | 0.134 | 31 | 0.137 | 16 | 2% |
| Asakusa | 0.041 | 79 | 0.051 | 54 | 22% |
| Hamacho | 0.024 | 162 | 0.029 | 88 | 24% |
| Honjoazumabashi | 0.020 | 175 | 0.030 | 86 | 45% |
| Morishita | 0.149 | 11 | 0.151 | 8 | 1% |
| Higashi-ojima | 0.024 | 166 | 0.029 | 87 | 25% |

| Station Name | JR + Subway | | JR + Subway + WB | | Centrality |
| | Centrality | Rank | Centrality | Rank | Increase |
| --- | --- | --- | --- | --- | --- |
| Etchujima | 0.025 | 203 | 0.031 | 191 | 25% |
| Ryogoku | 0.115 | 9 | 0.118 | 8 | 3% |
| Asakusa | 0.041 | 140 | 0.047 | 131 | 14% |
| Hamacho | 0.016 | 353 | 0.021 | 253 | 28% |
| Honjoazumabashi | 0.015 | 370 | 0.020 | 260 | 36% |
| Morishita | 0.065 | 78 | 0.068 | 74 | 5% |
| Higashi-ojima | 0.016 | 357 | 0.021 | 252 | 28% |

## 5  Conclusion

This paper investigated five transport networks and their combination networks in terms of complex network. The following implications have been revealed from this study on transport networks in Japan.; (1) The global efficiency of weighted network in Euclidean space becomes large and their values fall within 0.7-0.8 and it doesn't depend on the network size. However, the local efficiency is small. It may imply that our transport network achieves the high global efficiency by sacrificing the local efficiency meaning redundancy of networks.; (2) it is found that the role of waterbus with river system in Tokyo is to contribute to increasing the robustness and centrality around stations to which waterbus connects.

The following research should be pursued in the near future. First is considering OD demand between two stations $i$ and $j$ as new weight. Although the weight of networks in this paper is recognized as distance on path, actually it is necessary to use traffic amount on path or OD (Origin Destination) demand on stations as weight to evaluate transport networks. Although the analysis and results aren't represented in this paper due to the limitation of available data, the efficiency represented by Eq. (2.1) will be replaced with the following equation to incorporate the OD demand.

$$E(\mathbf{G}) = \sum_{i \neq j \in \mathbf{G}} \varepsilon_{ij} D_{ij} \qquad (5.1)$$

Where, $D_{ij}$ stands for the ratio of traffic amount between station $i$ and $j$ to total traffic amount. Second is finding algorithms to construct optimized waterbus or bus line networks with measures associated with the complex network. We believe that the architecture and feature found in the complex network will be utilized in bus line generation algorithms and give us useful information.

## Acknowledgements

This study is supported by the Ministry of Education, Culture, Sports, Science and Technology, Grand-in-Aid for Scientific Research (B), 17360424, 2005. We thank Kazue Honda for contribution to collecting station data of railways and subways.

## References

1. Watts, D.J., Strogatz, S.H. (1998) Collective Dynamics of 'Small-World' Networks, Nature, Vol.393, pp.440-442
2. Watts, D.J.: Small Worlds, Princeton University Press, (1999)
3. Strogatz, S.H. (2001) Exploring Complex Networks, Nature, Vol.410, pp.268-276,
4. Barabási, A.-L, Albert, R. (1999) Emergence of Scaling in Random Networks, Science, Vol.286, pp.509-512
5. Gastner, M.T., Newman, M.E.J. (2004) Shape and Efficiency in Spatial Distribution Networks, arXiv:cond-mat/0409702
6. Gastner, M.T., Newman, M.E.J. (2004) The Spatial Structure of Networks, arXiv.cond-mat/0407680
7. Kaiser, M., Hilgetag, C.C. (2004) Spatial Growth of Real-world Networks, Phys. Rev. E 69, 036103

8. Latora, V., Marchiori, M. (2002) Is the Boston Subway a Small-World Network?, Physica A, Vol.314, pp.109-113
9. Latora, V., Marchiori, M. (2001) Efficient Behavior of Small-World Networks, Phys. Rev. lett. Vol.87, No.19
10. Sen, P., Dasgupta, S., Chatterjee, A., Sreeram, P.A., Mukherjee, G., Manna, S.S. (2003) Small-World Properties of the Indian Railway Network, Phys. Rev. E 67, 036106
11. Seaton, K.A., Hackett, L.M. (2004) Stations, Trains and Small-World Networks, Physica A, Vol.339, pp.635-644
12. Sienkiewicz, J., Holyst, J.A. (2005) Statistical Analysis of 22 Public Transport Networks in Poland, Phys. Rev. E 72, 046127
13. http://www.mapion.co.jp/
14. http://www.tokyometro.jp/e/index.html
15. http://www.jreast.co.jp/e/index.html
16. http://www.ktr.mlit.go.jp/kyoku/river/pavilion/ship_08.htm

# Network Design via Flow Optimization

Yusuke MATSUMURA[1], Hidenori KAWAMURA[2],
Koichi KURUMATANI[3], and Azuma OHUCHI[2]

[1] Graduate School of Information Science and Technology, Hokkaido University
   yusuke_m@complex.eng.hokudai.ac.jp
[2] Graduate School of Information Science and Technology, Hokkaido University,
   CREST, JST {kawamura, ohuchi}@complex.eng.hokudai.ac.jp
[3] National Institute of Advanced Industrial Science and Technology, CREST, JST
   k.kurumatani@aist.go.jp

**Summary.** In the network field, various researches have tackled optimizing flow in networks. As shown from previous researches, flow optimality is dependent on the link structure and the routing algorithm that controls the flow. Therefore, a research frame is desirable that optimizes each of them independently but still simultaneously handles both. This study proposes a new scheme called Optimal Flow Network Design Problem (OFNDP) and applies it to a queueing network system; flow optimization is also performed.

In the experiment, the number of traffics and each service rate were given to nodes and between each node respectively, and then the flow of each traffic was decided to minimize waiting time by a flow control algorithm based on the dijxtra method. Link structure was also optimized by a genetic algorithm under the constraint of a limited number of links. As a result, sub-optimal flow was obtained with short average waiting time for each traffic.

## 1 Introduction

When the flow of such traffic as the packet, the people, and vehicles is given to a graph constructed with nodes and links, the graph is called a network. In a network, traffic occurs on a certain node, moves the link, and arrives at the node by other nodes. Flow must be efficient under a particular evaluation in the network when assuming that it is a system whose purpose is to provide services. In the network field, various researches have tackled flow optimization in networks [7].

Up to now, in the telecommunication network field, research has been done on the design of the routing algorithm that controls the flow of traffic. The routing algorithm gives the route to the traffic that occurs between each node in consideration of the crowded condition of traffic in each node when the traffic processing performance of the nodes, the link structure, and the communication band are given. Research on an adaptive routing algorithm, which

Y. MATSUMURA et al.: *Network Design via Flow Optimization*, Studies in Computational Intelligence
(SCI) **56**, 115–128 (2007)
www.springerlink.com       © Springer-Verlag Berlin Heidelberg 2007

is redundant for active changes in the link structure due to the breakdown of nodes etc., has been especially activated in recent years. These researches have shown that optimality of the flow of traffic in the network depends on the link structure.

On the other hand, the authors previously modeled a telecommunication network on a queueing network and investigated the features of link structure that minimize traffic waiting time. In our previous research [11, 10], sub-optimal link structure differed depending on the service rate given to the nodes and the amount of traffic given between each node when the routing algorithm is given to the network. Moreover, if the features of the link structure are different, even if the given amount of traffic and service rate are the same, waiting time increases and decreases, too. These results argue that it is important to design the link structure of the network in consideration of each given condition.

As shown from previous researches, flow optimality is dependent on link structure and the routing algorithm that controls the flow. Therefore, the frame of the research is desirable that optimizes each of them independently but also simultaneously handles both. This study proposes an Optimal Flow Network Design Problem (OFNDP) for a telecommunication network that simultaneously optimizes the flow and link structure. In this paper, to investigate the relations between flow efficiency and link structures, we statically give some sets of parameters such as the communication band of each link and the service rate of each node, although they should be designed in telecommunication networks [7].

The goal of this study is to establish new network building schemes, such as building scheme of an optimal flow network with optimal flow control algorithm and link structure from scratch, and reform scheme of current networks. To accomplish this goal, we start this study from revealing an optimal pair of link structure and flow control algorithm for a network which has a type of traffic occurrence. We consider ignorable networks of linking costs, such as "P2P" and "Ad-hoc Wireless Networks" for the first step.

This paper is organized as follows. Section 2 defines OFNDP with the queueing system theory, and Section 3 describes the derivation methodology of optimal flow. Section 4 describes some quantitative indices for analysis of link structure, and Section 5 shows the results of our numerical experiments, which are further discussed in Section 6. Section 7 concludes the paper.

## 2 Optimal Flow Network Design Problem

In this paper, we propose a new problem named Optimal Flow Network Design Problem (OFNDP) that focuses on the importance of optimizing the flow of traffics in a network. OFNDP differs because it simultaneously optimizes the flow and the link structure [6, 7, 5]. The definition of OFNDP is as follows.

## 2.1 Definition

An OFNP network consists of $n$ nodes with a queueing system and provides a particular kind of service for traffic. Node $i$ has a static attribute $\mu_i$ that represents the service rate of the queueing system in node $i$. Here, traffic means the minimum unit of packets that flow in the network. Each pair of nodes $i, j$ has a static parameter $f_{ij}$ that represents the flow value, which is the number of traffics that occur on node $i$ and head to node $j$ per unit time.

Let us define an arc that allows traffics to pass from node $i$ to node $j$ as $(i, j)$. Each arc $(i, j)$ has a static parameter $b_{ij}$, which is the capacity of the limiting value of the total number of traffics that pass through arc $(i, j)$.

Then, an OFNP network is defined as directed graph $D = (\boldsymbol{V}, \boldsymbol{A})$ where $\boldsymbol{V}$ and $\boldsymbol{A}$ are sets of nodes and arcs, respectively. Here we set $\boldsymbol{A} = \{(i, j) | i, j \in \boldsymbol{V}, i \neq j\}$, and define $\boldsymbol{C}$ as a set of all cycles in $\boldsymbol{A}$. Here, $|\boldsymbol{V}| = n$ and $|\boldsymbol{A}| = n(n-1)$.

Let us define a flow variable $x_{ij}^{pq}$ that represents the total number of traffics that occur on node $p$, head to node $q$, and pass through arc $(i, j)$ per unit time. Thus, the total number of traffics that pass through arc $(i, j)$ per unit time represented by $y_{ij}$ and the arrival rate of node $i$ represented by $\lambda_i$ are derived as follows:

$$y_{ij} = \sum_{p,q \in \boldsymbol{V}} x_{ij}^{pq} \tag{1}$$

$$\lambda_i = \sum_{j \in \boldsymbol{V}} y_{ji}. \tag{2}$$

In a queueing network, the waiting time of traffic at each node can be derived analytically by selecting the traffic model of the queueing system such as M/M/S, M/G/S etc. In this paper, we selected a very general traffic model, M/M/1[4].

In an M/M/1 queueing network, the average amount of traffic retained by the queue in node $i$ is represented by $\rho_i$, supposing that the queue capacity of nodes is infinite. It is derived as follows [8]:

$$\rho_i = \frac{\lambda_i}{\mu_i - \lambda_i}. \tag{3}$$

Little's result [9, 14] shown below designates the relationship between $\tau_i$ and $\rho_i$, where $\tau_i$ is the average waiting time of each traffic at node $i$.

$$\rho_i = \lambda_i \tau_i \tag{4}$$

---

[4] The M/M/1 model is a queueing model where both the distribution of traffic arrivals and service times is assumed to be exponential, and there is a single server.

Thus,

$$\tau_i = \frac{1}{\mu_i - \lambda_i}. \tag{5}$$

## 2.2 Formulate

The main purpose of OFNDP is to derive the optimal flow and link structure represented by $\boldsymbol{E}$. $\boldsymbol{E}$ is formed in the course of the derivation of the optimal flow, as shown in Eq. (13), under a constraint of the number of links $m$, as shown in Eq. (7). In this paper, we evaluate flow optimality by average traffic waiting time, as shown in Eq. (12). When supposing a transportation network, the physical traveling time of traffic will be considered.

This problem derives optimal flow $x_{ij}^{pq}$ and link structure $\boldsymbol{E}$ by giving $\boldsymbol{V}, \mu_i, \boldsymbol{A}, f_{ij}, b_{ij}$, and $m$, as shown below.

$$\min_{x_{ij}^{pq}} f \tag{6}$$

subject to

$$|\boldsymbol{E}| \leq m \tag{7}$$

$$\mu_i > \lambda_i \quad (\forall i \in \boldsymbol{V}) \tag{8}$$

$$\sum_{i \in \boldsymbol{V}} x_{ij}^{pq} - \sum_{k \in \boldsymbol{V}} x_{jk}^{pq} = \begin{cases} -f_{pq} & j = p \\ 0 & j \neq p, q \\ f_{ij} & j = q \end{cases} \quad (\forall p, q \in \boldsymbol{V}) \tag{9}$$

$$0 \leq \sum_{p,q \in \boldsymbol{V}} x_{ij}^{pq} \leq b_{ij} \quad (\forall i, j \in \boldsymbol{V}) \tag{10}$$

$$\prod_{(i,j) \in \boldsymbol{c}} x_{ij}^{pq} = 0 \quad (\forall \boldsymbol{c} \in \boldsymbol{C}, \forall p, q \in \boldsymbol{V}) \tag{11}$$

where

$$f = \frac{1}{\sum_{p,q \in \boldsymbol{V}} f_{pq}} \sum_{i,j \in \boldsymbol{V}} y_{ij} \tau_j \tag{12}$$

$$\boldsymbol{E} = \{\{i, j\} | y_{ij} > 0\}. \tag{13}$$

# 3 Derivation Methodology of Optimal Flow

We therefore treat this problem and obtain sub-optimal solutions as follows: We set a flow control algorithm to the network that calculates the sub-optimal flow between two nodes using waiting time information $\tau_i$ of all nodes. Under this algorithm, we set a searching algorithm that searches all pair of nodes $i, j$ that make $y_{ij}$ and $y_{j,i}$ be 0. This searching algorithm helps create sub-optimal flow with the following constraint, as shown in Eq. (7), and the operations performed by this algorithm are equivalent to decide link structure $\boldsymbol{E}$.

Hereinafter, we describe a flow control algorithm with the dijxtra method and a searching algorithm of the link structure with a genetic algorithm.

### 3.1 Flow Control Algorithm

The algorithm is described as follows:

1. Get an arc $(p, q) \leftarrow \boldsymbol{A}$
2. Set $\lambda_i := 0$ for each nodes for initialization
3. Set $\tau_i := \frac{1}{\mu_i - \lambda_i}$ for each node.
4. Select a node $i$, and give a path with minimum waiting time for each pair of node $i, j$ where $i \neq j$ by the dijxtra method on the block, and assign flow $x_{ij}^{pq} = f_{pq}$.
5. Update each $\lambda_i$ by the procedure shown in Eq. (2)
6. Repeat the procedure from (1) to (5) till $\boldsymbol{A}$ becomes $\emptyset$.

### 3.2 Link Structure Searching Algorithm

Simple Genetic Algorithms (SGA) are used as searching algorithms for sub-optimal link structure. They are based on a general SGA procedure, as shown below. An adjacency matrix of link structure $\boldsymbol{E}$ is used as a gene expression of the individual, as shown in Figure 1. All initial individuals are generated with a random graph model.

The selection is made by the roulette selection and elite preservation is applied. Mutation is operated as follows. First, one link is removed at random, and then to keep the previous number of links, a link is given to a couple of nodes without links.

Crossover for individual $i$ is executed as follows. First, we randomly select individual $j$ as the target individual; then, we randomly select number $k$ as the target node's number. The $k$th row in the adjacency matrix for individuals $i$ and $j$ is removed as chromosomes $p_i$ and $p_j$, respectively, and a different gene for each column between $p_i$ and $p_j$ is removed, and $p_i^*$ and $p_j^*$ are obtained as a new chromosome. The point of division of $p_i^*$ and $p_j^*$ is decided at random, and intersection is performed. If $p_i^*$ doesn't retain the number of links before

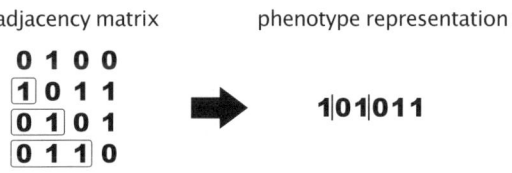

**Fig. 1.** An example of the phenotype representation of the adjacency matrix. Each element except for diagonal elements is taken out from the lower triangle matrices

the intersection, it is repeated to reverse the value of the randomly selected gene in $p_i^*$ until it takes back the previous number of links. Let $p_i^*$ be returned to $p_i$, and intersection for individual $i$ is performed.

# 4 Quantitative Indices of Complex Networks

In recent researches on complex networks, various quantitative indices have been used to analyze link structure. We use quantitative indices in the following [1, 3, 13]. In this paper, each vertex in the following means a node.

## 4.1 Clustering Coefficient

Clustering coefficient $C_i$ for vertex $v_i$ is the proportion of links between vertices within its neighborhood divided by the number of links that might exist between them. It is used extensively for average clustering coefficient $\bar{C}$. In addition, the range of each $C_i$ and $\bar{C}$ is 0 to 1.0.

## 4.2 Shortest Path Length

The shortest path length for a pair of vertices $(i, j)$ is the sum of the edge length within the shortest path of $(i, j)$. It is used extensively for average shortest path length $L$.

## 4.3 Betweenness Centrality

Betweenness centrality $B_i$ for vertex $v_i$ is the number of shortest paths that include vertex $v_i$ for each pair of vertices. Vertices that occur on many of the shortest paths between other vertices have higher betweenness than those that do not. It is used extensively for average betweenness $\bar{B}$.

## 4.4 Assortativity Coefficient

Assortativity coefficient is an index that shows the uniting correlation of nodes in a graph expressed by $r$. When high degree vertices tend to be connected to high degree vertices, the assortativity coefficient is positive, which means the graph is assortative, and vice versa. In addition, the range of $r$ is $-1 \leq r \leq 1$, and here, $r = 1$ means completely assortative, $r = -1$ means completely disassortative, and $r = 0$ means neutral, i.e., random.

**Table 1.** Experimental Parameters

| | 1 | 2 | | 3 | |
|---|---|---|---|---|---|
| | $j = \{1,\ldots,100\}$ | $j = \{1,\ldots,5\}$ $j = \{6,\ldots,100\}$ | | $j = \{1,2,3\}$ $j = \{4,\ldots,100\}$ | |
| $n$ | | 100 | | | |
| $k$ | | $4 \leq k \leq 99$ $(k = 2 \cdot m/n)$ | | | |
| $f_{ij}$ | 1 | 15 | 1 | 1 | 1 |
| $b_{ij}$ | | $\infty$ | | | |
| $\mu_j$ | 3,500 | 3,500 | 3,500 | 35,000 | 3,500 |

# 5 Numerical Experiments

To investigate relationships between link structure and traffic flow, the following experiments were performed that assumed three situations in the telecommunication network.

In each situation, experimental parameters that reflect a certain situation are set to OFNDP, and topology search is performed. Since the average waiting time of traffic is compared to one, fitness is generated with a random graph, a small-world network, and a scale-free network. The characteristics of each obtained topology are also compared. Here, we obtained random, small-world and scale-free link structures by Erdos model [4], WS model [12], and BA model [2], respectively.

In each experiment, we give $n$ nodes and perform topology search and characteristic analysis for each *bark*, which is the average degree of nodes with a range of $2(n-1)/n \leq \bar{k} \leq n-1$. Here, *bark* designates link density on the community given by $\bar{k} = n \cdot m/2$, where $m$ is the number of links.

## 5.1 Problem Setting

We assume the following three sets of parameters:

1. Each pair of nodes communicates a uniform amount and the service rates of all nodes are uniform.
2. Based on (1), but the amount of the traffic that departs for several nodes is greater than for others.
3. Based on (1), the service rates of several nodes are higher than that of some others.

Setting (1) represents a situation in which all nodes broadcast all information to each other at all times. Setting (2) represents a situation in which several nodes mainly receive information from other nodes, but all nodes communicate some information to each other. And setting (3) represents a situation identical to (1), but several nodes have high processing ability.

The experimental parameters for these three settings are shown in Table 1. Due to the restriction shown in Eq. (8), it is better to give a high value with more than enough $n$ to service rate $\mu$; however, it is not essential.

## 5.2 Results

We obtained optimized link structures for each setting by performing topology searches. The SGA parameters – generation, number of individuals, mutation probability, and crossover probability – were set to 10,000, 30, 10%, and 50%, respectively. We carried out 10 trials for each setting, and all results are shown as averages of 10 trials. Figures 2(a), 2(b), and 2(c) show the average traffic waiting times for each setting. The X axis indicates average degree $\bar{k}$, and the Y axis indicates the average waiting time calculated by Eq. (12). The random, ba, ws($\alpha$), and ga indices correspond to the topology generation methods, where $\alpha$ in ws is the exchange probability of the link. For each setting, optimized topology shortened the average waiting time for each $\bar{k}$, and as the average degree increased, waiting time became shorter. In particular, the average waiting time of the obtained topology for setting (3) was much shorter than the others.

Figure 3(a) shows averaged clustering coefficient $\bar{C}$ and averaged shortest path length $L$ generated by random, ba, and ws for each setting. Figure 3(b) shows the same indices for the optimized link structures. The X axis indicates $\bar{k}$, the Y-LEFT axis indicates $\bar{C}$ corresponding to the lines, and the Y-RIGHT axis indicates $L$ corresponding to the points. Each description (S$x$) in the indexes means that "setting ($x$)." For setting (1), optimized topology has a smaller $\bar{C}$ than random and a shorter $L$ than random. Moreover, it has similar $L$ to ba in the range of $\bar{k} < 20$. For setting (2), optimized topology has similar characteristics to on in setting (1) in the range of $\bar{k} < 15$, and it only has higher $C$ near $\bar{k} = 25$. For setting (3), optimized topology has higher $C$ for smaller $\bar{k}$ near $L = 2$ in the range of $\bar{k} < 20$.

Figure 4(a) shows averaged betweenness centrality $\bar{B}$ and the standard deviation of betweenness centrality $s(B)$ of the link structures generated by random, ba, and ws for each setting. Figure 4(b) shows the same indices for the optimized link structures. The X axis indicates $\bar{k}$, the Y-LEFT axis indicates $\bar{B}$ corresponding to the lines and $b$ in the indexes, and the Y-RIGHT axis indicates $s(B)$ corresponding to the points and $bdev$. Each description (S$x$) in the indexes means that "setting (x)." For setting (1), optimized topology has a similar $\bar{B}$ to random in the full range of $\bar{k}$ and a slightly larger $s(B)$ in the range of $\bar{k} < 12$. For setting (2), optimized topology has similar characteristics to (1), and it only has a larger $s(B)$ near $\bar{k} = 25$. For setting (3), optimized topology has a smaller $\bar{B}$ than the others in the range of $\bar{k} < 20$ and a larger $s(B)$ than the others in the full range of $\bar{k}$.

Figure 5(a) shows the standard deviation of degree $s(k)$ and assortativity coefficient $r$ of the link structures generated by random, ba, and ws for each setting. Figure 5(b) shows the same indices for the optimized link structures. The X axis indicates $\bar{k}$, the Y-LEFT axis indicates $s(k)$ corresponding to the lines and $kdev$ in the indexes, and the Y-RIGHT axis indicates $r$ corresponding to the points and $r$ in the indexes. Each description (S$x$) in the indexes means "setting (x)." For setting (1), optimized topology has a similar $s(k)$ and $r$ to

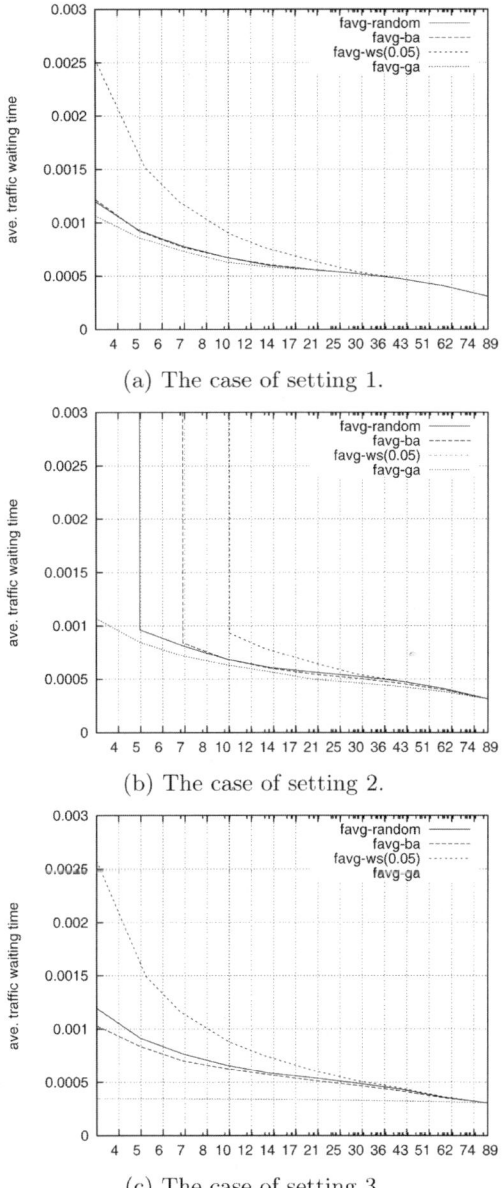

(a) The case of setting 1.

(b) The case of setting 2.

(c) The case of setting 3.

.  . The obtained results of average waiting time of traffic. In each figure, the X axis indicates the average degree ¯, and the Y axis indicates the average waiting time. The notations of "random", "ba", "ws( )", and "ga" correspond to the topology generation methods, where    in ws is the exchange probability of the link.

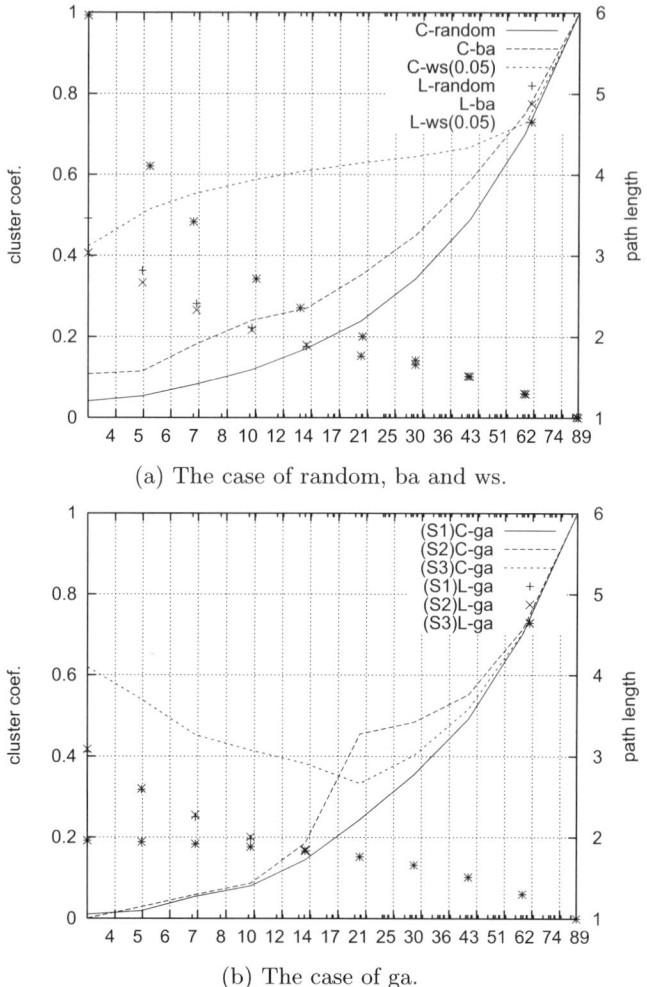

(a) The case of random, ba and ws.

(b) The case of ga.

**Fig. 3.** The results of the average clustering coefficient and the average shortest path length. The X axis indicates the average degree $\bar{k}$, the Y-LEFT axis indicates the average clustering coefficient $\bar{C}$ corresponding to the lines, and the Y-RIGHT axis indicates the average shortest path length $L$ corresponding to the points. Each description (S$x$) in the indexes means that "setting ($x$)"

random in the full range of $\bar{k}$. For setting (2), optimized topology has similar characteristics to on in setting (1), and it only has more higher $s(k)$ and lower $r$ than the others near $\bar{k} = 25$. For setting (3), optimized topology has higher $s(k)$ than the others in the full range of $\bar{k}$ and lower $r$ than the others in the range of $\bar{k} < 7$.

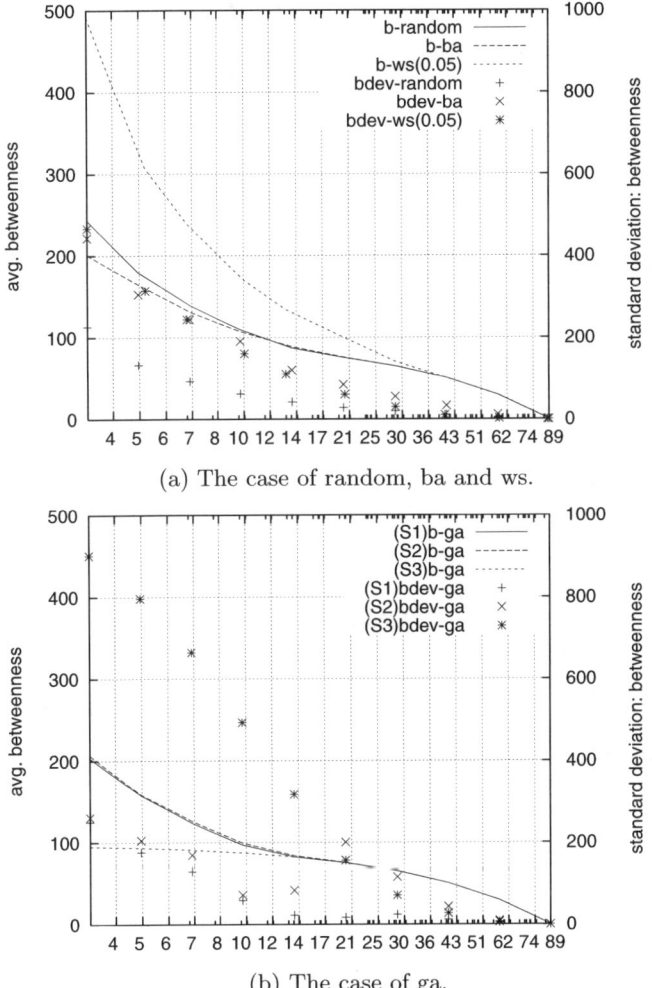

(a) The case of random, ba and ws.

(b) The case of ga.

**Fig. 4.** The results of the average betweenness and the standard deviation of betweenness. The X axis indicates the average degree $\bar{k}$, the Y-LEFT axis indicates the average betweenness $\bar{B}$ corresponding to the lines, and the Y-RIGHT axis indicates the standard deviation of betweenness $s(b)$ corresponding to the points. Each description (S$x$) in the indexes means that "setting ($x$)"

## 6 Discussion

In setting (1), the fitness of optimized topology is higher than the others when $\bar{k}$ is smaller than 20, and its characteristics are synthetically similar to random. The most characteristic index is $\bar{C}$, which is smaller than the others. We assume that this limited number of links would be better used for

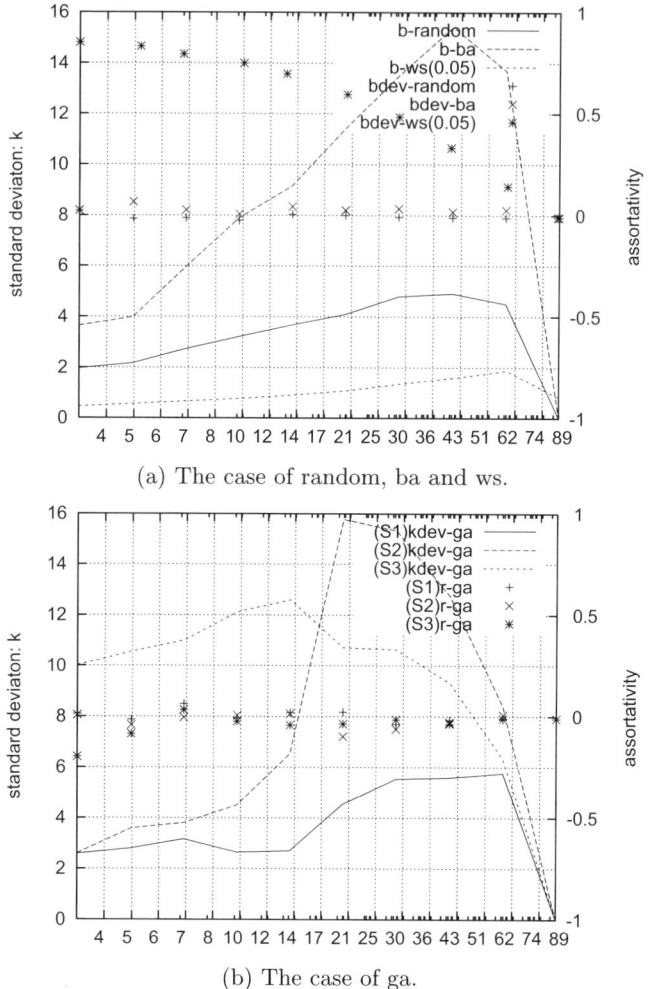

(a) The case of random, ba and ws.

(b) The case of ga.

**Fig. 5.** The results of the standard deviation of degree and the assortativity coefficient. The X axis indicates the average degree $\bar{k}$, the Y-LEFT axis indicates the standard deviation of degree $s(k)$ corresponding to the lines, and the Y-RIGHT axis indicates the assortativity coefficient $r$ corresponding to the points. Each description (S$x$) in the indexes means that "setting ($x$)"

the connection between clusters than for increasing the closeness of a certain cluster; therefore, $L$ is shortened. This explains the low fitness of small world. It also explains that it is unfit for this situation of unique nodes, e.g., the hub and the shortcut, which exist in scale-free and small-world. For this setting, it is desirable to design a topology to evenly connect nodes and without forming any specific cluster.

In setting (2), the fitness of optimized topology is higher than the others regardless of $\bar{k}$, and its characteristics are synthetically similar to random. However, some indices differ greatly from others near $\bar{k} = 25$, such as high $\bar{C}$, the large deviation of betweenness, the large deviation of degree, and high disassortativity. We assume that enough links allow all nodes to connect to nodes that receive heavy traffic because mass traffic would rather be sent without passing any nodes. It follows that such nodes have many links, and the others have few links. Therefore, high cluster coefficient, large deviation of betweenness, and high disassortativity are observed. For such a setting, it is necessary to consider a design strategy based on the available number of links.

In setting (3), the fitness of optimized topology is much higher than the others regardless of $\bar{k}$, and characteristic indices are found where $\bar{k}$ is less than 20, such as high $\bar{C}$, short $L$, which is smaller than 2, and the large deviation of betweenness. In particular, disassortativity is very high where $\bar{k}$ is smaller than 7. Due to short $L$, we assume that most nodes have links to high capacity nodes; therefore, traffic can be sent in two hops, which also explains high $\bar{C}$. Moreover, it explains the large deviation of degrees and betweenness because almost all links are monopolized by high capacity nodes. However, disassortativity is not high where $\bar{k}$ is larger than 7; so there are enough links in the network to facilitate no waiting and high speed communication. For such a setting, we can reduce the number of links based on the diverseness of the performance of nodes.

# 7 Conclusions

This research modeled a telecommunication network on a queueing network and proposed an Optimal Flow Network Design Problem. In the experiment, the amount of traffic and each service rate were given to the nodes and between each node respectively, and then the flow of traffic was determined to minimize the waiting time of traffic by a flow control algorithm based on the dijxtra method. Link structure was also optimized by a genetic algorithm under a constraint of the limited number of links. As a result, sub-optimal flow was obtained with short average waiting time for traffic.

Obtained link structures were analyzed using the complex network theory, and it became clear that sub-optimal link structures differ depending on the amount of traffic and the service rate. The difference at waiting time in a random graph and a free scale and the one in the sub-optimal link structure has especially grown by restricting a few links. This result suggests that we should especially consider link structure when a great number of links needs much cost to generate.

Actually, flow is often controlled from such local information as the waiting time of the neighborhood node, although the flow control algorithm based on the dijxtra method used in the experiment helps make flows sub-optimal in a condition that allows the algorithm to know all the information about the waiting time at each node. In the future we want to examine a routing algorithm that can make sub-optimal flow only using local information by analyzing the amount of traffic allocated in each link. We will also examine a link generation algorithm that creates sub-optimal link structure under such situations as the number of nodes increasing.

## Acknowlegement

We wish to thank Dr. Masahito Yamamoto and Dr. Masashi Furukawa, Hokkaido University, Dr. Akira Namatame, National Diffence Academy, Dr. Satoshi Kurihara, Osaka University, Dr. Tomohisa Yamashita, Dr. Kiyoshi Izumi, National Institute of Advanced Industrial Science and Technology, and the anonymous reviewers for their considerable cooperation and valuable advices. This research was partially supported by a Grant-in-Aid of CREST of the Japan Science and Technology Agency.

## References

1. R. Albert and A.-L. Barabási. Statistical mechanizm of complex network. *Review of Modern Physics*, 74, 2002.
2. A.-L. Barabási and R. Albert. Emergence of scaling in random networks. *Science*, 286, 1999.
3. S. N. Dorogovtsev and J. F. F. Mendes. Evolution of networks. *Advances in Physics*, 51(4), 2002.
4. P. Erdós and A. Rényi. On random graphs. *Publicationes Mathematicae*, 6, 1959.
5. R. E. Gomory and T. C. Hu. Synthesis of a communication network. *J. SIAM*, 12(2), 1961.
6. T. C. Hu. Multi-commodity network flows. *J. ORSA*, 15(2), 1967.
7. T. Kenyon. *High-performance data network disign: design techniques and tools.* Butterworth-Heinemann, 2002.
8. L. Kleinrock. *Queueing systems*, volume 1. Wiley-Interscience, 1975.
9. J. Little. A proof of the queueing formula $l = \lambda w$. *Operetions Resesearch*, 9, 1961.
10. Y. Matsumura, H. Kawamura, and A. Ohuchi. Desirable design methodology for queueing networks. *In Proc. of the First International Workshop on Artificial Computational Economics & Social Simulation 2005*, 2005.
11. Y. Matsumura, H. Kawamura, and A. Ohuchi. Desirable design of queueing networks excluding linking costs. *In Proc. of the Workshop on Emergent Intelligence on Networked Agents (WEIN' 06)*, 2006.
12. S. H. Strogatz. Exploring complex networks. *Nature*, 410, 2001.
13. D. J. Watts. *Nature*, 393(440), 1998.
14. W. Whitt. A review of $l = \lambda w$ and extensions. *Queueing Systems*, 9(3), 1991.

# Gibbs measures for the network

Syuji Miyazaki

Graduate School of Informatics, Kyoto University, Kyoto 606-8501, JAPAN
syuji@acs.i.kyoto-u.ac.jp

## 1 Introduction

One of the most remarkable points about deterministic chaos is the duality
consisting of irregular dynamics and fractal structure of the attractor in the
phase space. Amplifying this relation between dynamics and geometry, we
will try to construct dynamics corresponding to a directed network structure such as the WWW. A directed network or graph can be represented
by a transition matrix. On the other hand, temporal evolution of a chaotic
piecewise-linear one-dimensional map with Markov partition can be governed
by a Frobenius-Perron matrix. Both transition matrices and Frobenius-Perron
matrices belong to a class of transition matrices sharing the same mathematical properties. The maximum eigenvalue is equal to unity. The corresponding
eigenvector is always a real vector, and evaluates the probability density of
visiting a subinterval of the map or a site of the network, which is commer-
cially valuable information in the field of the WWW [1]. Relating these two
matrices to each other, we are able to represent the structure itself of the
directed network as a dynamical system. Once we relate the directed network
to chaotic dynamics, several approaches to deterministic chaos can be also
applied to graph theory.

In chaotic dynamical systems, local expansion rates which evaluate an or-
bital instability fluctuate largely in time, reflecting a complex structure in the
phase space. Its average is called the Lyapunov exponent, whose positive sign
is a practical criterion of chaos. There exist numerous investigations based on
large deviation statistics in which one considers distributions of coarse-grained
expansion rates (finite-time Lyapunov exponent) in order to extract large deviations caused by non-hyperbolicities or long correlations in the vicinity of
bifurcation points [2].

In general, statistical structure functions consisting of weighted averages,
variances, and these partition functions as well as fluctuation spectra of coarse-
grained dynamic variables can be obtained by processing the time series nu-
merically. In the case of the piecewise-linear map with Markov partition, we

S. Miyazaki: *Gibbs measures for the network,* Studies in Computational Intelligence (SCI) **56**,
129–137 (2007)
www.springerlink.com

can obtain these structure functions analytically. This is one of the reasons why we correspond a directed network to a piecewise-linear map. We herein try to apply an approach based on large deviation statistics in the research field of chaotic dynamical systems to network analyses. We discuss mainly the Gibbs measure in the present paper.

## 2 One-Dimensional Map Corresponding to Directed Network

Let us define this adjacency matrix $A$, where $A_{ij}$ is equal to unity if the node $j$ is linked to $i$. If not, $A_{ij}$ is equal to zero. In the case of undirected network, $A_{ij} = A_{ji} = 1$ holds if the nodes $i$ and $j$ are linked to each other. If not, $A_{ij}$ is equal to zero. Transition matrix $H$ can be derived straightforwardly from the adjacency matrix. The element $H_{ij}$ is equal to $A_{ij}$ divided by the number of nonzero elements of column $j$. We will consider the very simple example whose transition matrix is given by

$$H = \begin{pmatrix} 0 & \frac{1}{2} \\ 1 & \frac{1}{2} \end{pmatrix}. \tag{1}$$

The maximum eigenvalue is always equal to unity. The right eigenvector $h$ with $Hh = h$ measures site importance in the context of the web network[1]. The left eigenvector $v$ with $vH = v$ satisfies $v_1 = v_2$, since all the nonzero elements are equal in the same column of $H$. This example corresponds to a directed network consisting of two nodes in which node 1 is always directed to node 2, and node 2 to both with equal probability. In other words, this is a special coin tossing in which heads (node 1) are always followed by tails (node 2), tails by heads or tails with equal probability.

This directed network can be represented by a map $f$ from unit interval $I = [0, 1)$ to $I$. We divide $I$ into $I_1$ and $I_2$ with $I = I_1 \cup I_2$ and $I_1 \cap I_2 = \phi$. The most simple choice of the map satisfying $f(I_1) = I_2$, $f(I_2) = I_1 \cup I_2$ is piecewise linear, and is given by $f(x) = x + 1/2$ $(0 \le x < 1/2)$, $2x - 1$ $(1/2 \le x \le 1)$ with $I_1 = [0, 1/2)$ and $I_2 = [1/2, 1)$. This is a simple example of Markov partition. The slope $f'(I_1) = 1$ is equal to output degree of node 1, $f'(I_2) = 2$ to that of node 2. Note that the slope of the map is equal to the output degree of the graph in this way and that the element of $H$ is equal to inverse of a slope or of an output degree.

In the case of chaotic dynamics caused by a one-dimensional map $f$, the trajectory is given by iteration. Its distribution at time $n$, $\rho_n(x)$, is given by the average of this delta function $\langle \delta(x_n - x) \rangle$, where $\langle \cdots \rangle$ denotes the average with respect to initial points $x_0$. The temporal evolution of $\rho$ is given by the following relation $\rho_{n+1}(x) = \int_0^1 \delta(f(y) - x)\rho_n(y)\,dy \equiv \mathcal{H}\rho_n(x)$. This operator $\mathcal{H}$, called the Frobenius-Perron operator, is explicitly given as

$\mathcal{H}G(x) = \sum_j \dfrac{G(y_j)}{|f'(y_j)|}$, where the sum is taken over all solutions $y_j(x)$ satisfying $f(y_j) = x$.

In the case of a piecewise-linear map with Markov partition, invariant density is constant for each interval. Taking these 2 functions as a basis, we can represent the Frobenius-Perron operator as this 2 by 2 matrix $H$. This is nothing but the transition matrix of the directed graph consisting of 2 nodes mentioned before. For an arbitrary transition matrix, all the elements are summed up to unity along the same column. For a Frobenius-Perron matrix corresponding to a piecewise linear map, this sum rule is also satisfied, and furthermore, all the nonzero elements must be equal in the same column. In this sense, our Frobenius-Perron matrix belongs to a subset of the whole transition matrix (stochastic matrix).

Right and left eigenvectors corresponding to eigenvalue 1 of the aforementioned Frobenius-Perron matrix $H$ are determined. The left eigenvector $v$ is given by $v = (1/2, 1/2)$, where it is so normalized that the sum of all elements is equal to unity. Note that the element is equal to the width of the subintervals $I_1$ and $I_2$ of the Markov partition. The right eigenvector $h$ gives the probability density to visit each subinterval, and is equal to $h = (4/3, 2/3)$, where it is so normalized that the inner product of the right and the left eigenvectors is equal to unity.

The Lyapunov exponent of the one-dimensional map is an average of the logarithm of the slope of the map with respect to its invariant density. Comparing the directed network, the map $f$, the matrix $H$, we find that the output degree of node $k$ is equal to the slope of the interval $I_k$. Thus the Lyapunov exponent of the network is found to be an average of the logarithm of the output degree of each node. This exponent quantifies the complexity of link relations. Degree distribution, a function describing the total number of vertices in a graph with a given degree (number of connections to other vertices), is often used to characterize link relations. The network Lyapunov exponent and its fluctuation are also supposed to be useful in the context of the network. In this way, we can relate the network structure itself to a chaotic dynamical system, and we try to characterize the network based on an approach to deterministic chaos, namely large deviation statistics, in other words, thermodynamical formalism. Large deviation statistics of Lyapunov exponents of chaotic dynamical systems is intensively discussed in the literature [2].

## 3 Large Deviation Statistics

Let us briefly describe large deviation statistics following the series of studies by Fujisaka and his coworkers [3, 4]. Consider a stationary time series of a dynamic variable $\tilde{u}\{t\}$ at time $t$. The average over time interval $T$ is given by this formula, $\overline{u}_T(t) = \dfrac{1}{T} \displaystyle\int_t^{t+T} \tilde{u}\{s\}\,ds$, which distributes when $T$ is finite.

When $T$ is much larger than the correlation time of $\tilde{u}$, the distribution $P_T(u)$ of coarse-grained $u = \bar{u}_T$ is assumed to be an exponential form $P_T(u) \propto e^{-S(u)T}$. Here we can introduce fluctuation $S(u)$ as $S(u) = -\lim_{T \to \infty} \frac{1}{T} \log P_T(u)$. When $T$ is comparable to the correlation time, correlation can not be ignored, so non-exponential or non-extensive statistics will be a problem, but here we do not discuss this point any further. Let $q$ be a real parameter. We introduce the generating function $M_q$ of $T$ by this definition: $M_q(T) \equiv \langle e^{qT\bar{u}_T} \rangle = \int_{-\infty}^{\infty} P_T(u)e^{qTu}\,du$. We can also here assume the exponential distribution and introduce characteristic function $\phi(q)$ as $\phi(q) = \lim_{T \to \infty} \frac{1}{T} \log M_q(T)$. The Legendre transform holds between fluctuation spectrum $S(u)$ and characteristic function $\phi(q)$, which is obtained from saddle-point calculations. $\frac{dS(u)}{du} = q, \quad \phi(q) = -S(u(q)) + qu(q)$. In this transform a derivative $d\phi/dq$ appears, and is a weighted average of $\bar{u}_T$, $u(q) = \frac{d\phi(q)}{dq} = \lim_{T \to \infty} \frac{\bar{u}_T e^{qT\bar{u}_T}}{M_q(T)}$, so we find that $q$ is a kind of weight index. We can also introduce susceptibility $\chi(q) = \frac{du(q)}{dq}$ as a weighted variance. These statistical structure functions $S(u), \phi(q), u(q), \chi(q)$ constitute the framework of statistical thermodynamics of temporal fluctuation, which characterize the static properties of chaotic dynamics. In order to consider dynamic properties, we can introduce this generalized spectrum density as a weighted average of conventional spectrum density as $I_q(\omega) = \lim_{T \to \infty} \left\langle \left| \int_0^T [\tilde{u}\{t+s\} - u(q)]e^{-i\omega s}ds \right|^2 e^{qT\bar{u}_T} \right\rangle / (T\, M_q(T))$. In the same way, the generalized double time correlation function is $C_q(t) = \lim_{T \to \infty} \lim_{\tau \to \infty} \langle (\tilde{u}\{t+\tau\} - u(q))(\tilde{u}\{\tau\} - u(q))e^{qT\bar{u}_T} \rangle / M_q(T)$. The relation between the two is given by the Wiener-Khintchine theorem: $C_q(t) = \int_{-\infty}^{\infty} I_q(\omega)\, e^{-i\omega t}d\omega/(2\pi)$ and $I_q(\omega) = \sum_{t=-\infty}^{\infty} C_q(t)e^{i\omega t}$. In the case of chaotic dynamics, $q$ of thermodynamical formalism is merely a weight index to process time series. This index $q$ can be used to control traffic in the context of the web network.

Let us consider the case of a one-dimensional map. Let $\tilde{u}[x_n]$ be a unique function of $x$, which is governed by the map $x_{n+1} = f(x_n)$. The question is how to obtain statistical structure functions and generalized spectral densities of $u$. The answer is to solve the eigenvalue problems of a generalized Frobenius-Perron operator. As we mentioned before, the characteristic function $\phi(q)$ is given by the asymptotic form of the generating function $M_q(n)$ in the limit of $n \to \infty$ corresponding to the temporal coarse-grained quantity

$\bar{u}_n = \dfrac{1}{n} \sum_{j=0}^{n-1} \tilde{u}[x_{j+m}]$, where we assume an exponential fast decay of time correlations of $u$. A generating function can be expressed in terms of invariant density,

$$M_q(n) = \int \rho_\infty(x) \exp\left[q \sum_{j=0}^{n-1} \tilde{u}[f^j(x)]\right] dx = \int \mathcal{H}_q^n \rho_\infty(x) dx,$$

where the generalized Frobenius-Perron operator $\mathcal{H}_q$ is defined and related to the original one as

$$\mathcal{H}_q G(x) = \mathcal{H}[e^{q\tilde{u}[x]} G(x)] = \sum_k \frac{e^{q\tilde{u}[y_k]} G(y_k)}{|f'(y_k)|}$$

for an arbitrary function $G(x)$ ($\mathcal{H}_0 = \mathcal{H}$). To obtain the above equation, the following relation is repeatedly used:

$$\mathcal{H}\left\{G(x) \exp\left[q \sum_{j=0}^{m} \tilde{u}[f^j(x)]\right]\right\} = (\mathcal{H}_q G(x)) \exp\left[q \sum_{j=0}^{m-1} \tilde{u}[f^j(x)]\right].$$

The normal Frobenius-Perron operator $\mathcal{H}$ depends on the map $f$ only. The generalized one $\mathcal{H}_q$ depends also on a dynamic variable $u$ and determines statistical structure functions and generalized spectral densities of $u$. For example, in the case of local expansion rates $\tilde{u}[x] = \log|f'(x)|$ whose average is the Lyapunov exponent, the generalized operator is explicitly given by $\mathcal{H}_q G(x) = \sum_k \dfrac{G(y_k)}{|f'(y_k)|^{1-q}}$.

In the case of the triangular network mentioned earlier, three subintervals constitute the Markov partition, such that $\mathcal{H}_q$ can be represented by a $2 \times 2$ matrix as $H_q = \begin{pmatrix} 0 & \frac{1}{2} \\ 1 & \frac{1}{2} \end{pmatrix} \begin{pmatrix} e^{q\tilde{u}[I_1]} & 0 \\ 0 & e^{q\tilde{u}[I_2]} \end{pmatrix}$ in the same way as $\mathcal{H}$, where $\tilde{u}[I_k] = \tilde{u}[x]$ with $x \in I_k$. Let $\nu_q^{(0)}$ be the maximum eigenvalue of $H_q$. The statistical structure functions and the generalized spectral density are given by

$$\phi(q) = \log \nu_q^{(0)}, \quad u(q) = \frac{d\phi(q)}{dq}, \quad \chi(q) = \frac{du(q)}{dq}, \tag{2}$$

$$I_q(\omega) = \int v_{(0)}(x)[\tilde{u}[x] - u(q)][J_q(\omega) + J_q(-\omega) - 1][\tilde{u}[x] - u(q)]h^{(0)}(x) dx, \tag{3}$$

where $J_q(\omega) = 1/\left[1 - (e^{i\omega}/\nu_q^{(0)})H_q\right]$, $v^{(0)}(x)$ and $h^{(0)}(x)$ are respectively the left and right eigenfunctions corresponding to the maximum eigenvalue $\nu_q^{(0)}$ of $H_q$. The generalized double time correlation function is given by $C_q(t) = \int v^{(0)}(x)[\tilde{u}[x] - u(q)][H_q/\nu_q^{(0)}]^t[\tilde{u}[x] - u(q)]h^{(0)}(x) dx$.

## 4 Gibbs Measures

Let us analyze the triangular directed network based on the large deviation statistics. We consider an arbitrary dynamic variable and set $\tilde{u}(I_1) = u_1$ and $\tilde{u}(I_2) = u_2$. The generalized Frobenius-Perron matrix $H_q$ can be represented as $H_q = \begin{pmatrix} 0 & e^{qu_2}/2 \\ e^{qu_1} & e^{qu_2}/2 \end{pmatrix}$. The largest eigenvalue is given by

$$\nu_q^{(0)} = \frac{e^{qu_2}}{4} + \frac{e^{qu_2}}{4}\sqrt{1 + 8e^{q(u_1-u_2)}}, \tag{4}$$

whose right and left eigenvectors $\begin{pmatrix} h_1(q) \\ h_2(q) \end{pmatrix}$, $(v_1(q)\, v_2(q))$ are also given by

$$v_1(q) = \frac{e^{qu_1}}{e^{qu_1} + \nu_q^{(0)}}, \qquad v_2(q) = \frac{\nu_q^{(0)}}{e^{qu_1} + \nu_q^{(0)}}, \tag{5}$$

$$h_1(q) = \frac{e^{qu_2}(e^{qu_1} + \nu_q^{(0)})}{e^{q(u_1+u_2)} + 2[\nu_q^{(0)}]^2}, \qquad h_2(q) = \frac{2\nu_q^{(0)}(e^{qu_1} + \nu_q^{(0)})}{e^{q(u_1+u_2)} + 2[\nu_q^{(0)}]^2} \tag{6}$$

satisfying $v_1(q) + v_2(q) = 1$ and $h_1(q)v_1(q) + h_2(q)v_2(q) = 1$.

The characteristic function is given by $\phi(q) = \log \nu_q^{(0)}$. The weighted average $u(q) = \dfrac{d\phi(q)}{dq}$ is explicitly given by

$$u(q) = u_2 + \left(1 - \frac{1}{2\sqrt{1 + 8e^{q(u_1-u_2)}}}\right)(u_1 - u_2). \tag{7}$$

Its asymptotic behaviors depend on the sign of $u_2 - u_1$. For $q\,\mathrm{sgn}(u_2 - u_1) \to +\infty\,(-\infty)$, we have $u(q) \to \frac{u_1+u_2}{2}\,(u_1)$. The weighted variance $\chi(q) = \dfrac{du(q)}{dq}$ is explicitly given by

$$\chi(q) = \frac{2(u_1 - u_2)^2 e^{q(u_1-u_2)}}{(1 + 8e^{q(u_1-u_2)})^{3/2}}. \tag{8}$$

The fluctuation spectrum $S(u)$, the generalized correlation function $C_q(t)$ and the generalized power spectrum $I_q(\omega)$ can be straightforwardly obtained. These statistical structure functions is obtained for other simple networks [5].

In the thermodynamic formalism in the context of the mathematical formulation of equilibrium statistical mechanics [6] and the ergodic theory of dynamical systems [7, 8], the variational principle [9]

$$P_f(\varphi) = \sup(h_\mu(f) + \int \varphi\, d\mu) \tag{9}$$

is often used, where $\varphi$ denotes a piecewise continuous function, $h_\mu(f)$ the Kolmogorov-Sinai entropy and the supremum is attained at a measure $\mu_\varphi$

called Gibbs measure, which depends on chosen function $\varphi$. Hereafter we confine ourselves to hyperbolic one-dimensional dynamical systems. The topological pressure $P_f(\varphi)$ may admit different analytical branches corresponding to different local structures of the invariant set reflected by different Gibbs measure [7, 10, 11]. The topological pressure $P_f(\varphi)$ and the characteristic function $\phi(q)$ are identical if one identifies $\varphi = qu - \log|f'|$. The Gibbs measure corresponds to $h_1(q)v_1(q)$ and $h_2(q)v_2(q)$.

Note that $H_q/v_q^{(0)}$ has unit eigenvalue and corresponds to a conventional Frobenius-Perron matrix of the induced map $\tilde{f}$

$$\tilde{f}(x; q, u) = \frac{1-a}{a}x + a, \qquad \left(x \in \tilde{I}_1 = [0, a)\right), \tag{10}$$

$$\frac{1}{1-a}(x-1) + 1, \qquad \left(x \in \tilde{I}_2 = [a, 1)\right) \tag{11}$$

with $a = \dfrac{e^{qu_1}}{e^{qu_1} + v_q^{(0)}}$. Note that the same conditions $[0, 1) = \tilde{I}_1 \cup \tilde{I}_2$, $\tilde{I}_1 \cap \tilde{I}_2 = \phi$, $\tilde{f}(\tilde{I}_1) = \tilde{I}_2$ and $\tilde{f}(\tilde{I}_2) = \tilde{I}_1 \cup \tilde{I}_2$ on the Markov partition as the original dynamics $f$ hold.

Let us consider the concrete dynamic variable $u_1 = 0$ and $u_2 = 1$ as an example of the case of $u_1 < u_2$. The stationary probability density to visit node 1 (heads) and node 2 (tails) are respectively equal to $h_1(0) = 4/3$ and $h_2(0) = 2/3$. The stationary visiting frequencies are $h_1(0)v_1(0) = 2/3$ and $h_2(0)v_2(0) = 1/3$, which is used in the context of the WWW. For $q \to +\infty\,(-\infty)$, the weighted average $u(q) = v_1(q)u_1h_1(q) + v_2(q)u_2h_2(q)$ is equal to the largest (smallest) possible local average value, i. e., $u(+\infty) = 1$ and $u(-\infty) = 1/2$. The former and the latter correspond respectively to the unstable fixed point of $f$ yielding the time series of $\tilde{u} : 111\cdots$ with average 1 (only tails) and the unstable periodic point $010101\cdots$ with average $1/2$ (heads and tails alternate). In the limit $q \to +\infty$, $\tilde{I}_2$ approaches $[0, 1)$ and the width of $\tilde{I}_1$ shrinks, and $\tilde{f}(x; q, u) \to x$ for $x \in \tilde{I}_2$ which implies that all points in $[0, 1)$ are marginal fixed points in this limit. The second iteration $\tilde{f} \circ \tilde{f}$ is given by

$$\tilde{f} \circ \tilde{f}(x; q, u) = \frac{x}{a}, \qquad \left(x \in \tilde{I}_1 = [0, a)\right), \tag{12}$$

$$\frac{x-1}{a} + \frac{1-a+a^2}{a}, \qquad \left(x \in \tilde{I}_{21} = [a, 1-(1-a)^2)\right), \tag{13}$$

$$\frac{x-1}{(1-a)^2} + 1, \qquad \left(x \in \tilde{I}_{22} = [1-(1-a)^2, 1)\right) \tag{14}$$

In the limit $q \to -\infty$, $\tilde{I}_1$ approaches $[0, 1)$ and the widths of $\tilde{I}_{21}$ and $\tilde{I}_{22}$ shrink, and $\tilde{f}(x; q, u) \to x$ for $x \in \tilde{I}_1$ which implies that all points in $[0, 1)$ are marginal periodic points with period 2 in this limit. Note that the weighted average $u(q)$ is equal to the smallest (largest) possible local average value for $q \to +\infty\,(-\infty)$ and $u_1 > u_2$.

In this way different local structures of the invariant set are reflected by different Gibbs measures. For this dynamical system $f$, there exist different Gibbs measures. For a fixed Gibbs measures $(v_1(q)h_1(q), v_2(q)h_2(q))$, one can construct a induced dynamical system $\tilde{f}$ whose invariant measure (SRB measure) $(\tilde{v}_1(0)\tilde{h}_1(0), \tilde{v}_2(0)\tilde{h}_2(0))$ is identical to the above-chosen Gibbs measure $(v_1(q)h_1(q), v_2(q)h_2(q))$.

## 5 Possible Applications

In the context of WWW, one may choose a dynamic variable $\tilde{u}$ as $\tilde{u}(I_i) = 1$ if node (web site) $i$ contains a specific keyword; if not $\tilde{u}(I_i) = 0$. Gibbs measures $h_i(q)v_i(q)$ for large positive $q$ reflect only the web sites including the specific keyword and these closely related sites. For large negative $q$, Gibbs measures reflects only the sites NOT including the keyword and these closely related sites. Only keyword independent measure $h_i(0)v_i(0)$ is widely used to estimate the importance in the web network. We believe that Gibbs measures $h_i(q)v_i(q)$ are very promising as a measure of site importance depending on search terms or a filtering technique of harmful contents.

## Acknowledgements

This study was partially supported by the 21st Century COE Program "Center of Excellence for Research and Education on Complex Functional Mechanical Systems"at Kyoto University.

## References

1. L. Page, S. Brin, R. Motwani, T. Winograd, 'The PageRank Citation Ranking: Bringing Order to the Web', (1998), http://www-db.standford.edu/~backrub/pageranksub.ps
2. H. Mori and Y. Kuramoto, Dissipative structures and chaos, (Springer, Berlin, 1998).
3. H. Fujisaka and M. Inoue, Prog. Theor. Phys. **77**, 1334 (1987); Phys. Rev. A **39**, 1376 (1989).
4. H. Fujisaka and H. Shibata, Prog. Theor. Phys. **85**, 187 (1991).
5. Syuji Miyazaki, IPSJ Journal, **47**, 795 (2006) [in Japanese]; Springer LNAI series 4012 "New Frontiers in Artificial Intelligence: Proceeding of the 19th Annual Conferences of the Japanese Society for Artificial Intelligence" pp.261-270 (2006); Prog. Theor. Phys. Suppl. No.162, 147 (2006).
6. Ya. G. Sinai, Usp. Math. Nauk. **27**, 21 (1972);
   D. Ruelle, Am. J. Math. **98**, 619 (1976);
   D. Ruelle, Thermodynamical Formalism, Encyclopedia of Mathematics and its applications, Vol. 5 (Addison-Wesley, Reading, 1978);
   R. Bowen, Equilibrium States and the Ergodic Theory of Anosov Diffeomorphisms, Lecture Notes in Mathematics, Vol. 470, (Springer, Berlin, 1975).

7. H. Mori, H. Hata, T. Horita and T. Kobayashi, Prog. Theor. Phys. Suppl. No. 99, 1 (1989).

8. J. P. Eckmann and I. Procaccia, Phys. Rev. A **34**, 659 (1986);
   G. Paradin and A. Vulpiani, Phys. Rep. **156**, 147 (1987);
   P. Grassberger, R. Badii and A. Politi, J. Stat. Phys. **51**, 135 (1988);
   D.Bessis, G.Paradin, G. Turchetti and S. Valenti, J. Stat. Phys. **51**, 109 (1988).

9. P. Walters, An Introduction to Ergodic Theory (Springer, Berlin, 1982).

10. M. J. Feigenbaum, I. Procaccia and T. Tel, Phys. Rev. A **39**, 5359 (1989).

11. W. Just and H. Fujisaka, Physica D **64**, 98 (1993).

# Extracting Users' Interests of Web-watching Behaviors Based on Site-Keyword Graph

Tsuyoshi Murata[1] and Kota Saito[2]

[1] Department of Computer Science, Graduate School of Information Science and Engineering, Tokyo Institute of Technology, 2-12-1 Ookayama, Meguro-ku, Tokyo 152-8552, Japan `murata@cs.titech.ac.jp`

[2] Department of Human System Science, Graduate School of Decision Science and Technology, Tokyo Institute of Technology, 2-12-1 Ookayama, Meguro-ku, Tokyo 152-8552, Japan `k.saito@kuramae.ne.jp`

Analyzing users' Web log data and extracting their interests of Web-watching behaviors are important and challenging research topics of Web usage mining. Users visit their favorite sites and sometimes search new sites by performing keyword search on search engines. Users' Web-watching behaviors can be regarded as a graph since visited Web sites and entered search keywords are connected with each other in a time sequence. We call this graph as a site-keyword graph. This paper describes a method for clarifying users' interests based on an analysis of the site-keyword graph. The method is for extracting subgraphs representing users' routine visit from a site-keyword graph which is generated from augmented Web audience measurement data (Web log data). Experimental result shows that our new method succeeds in finding subgraphs which contain most of users' interested sites.

## 1 Introduction

Analyzing users' Web watching behaviors is one of the important and challenging research topics of Web usage mining. If users' interests can be automatically detected from users Web log data, the detected interests can be used for information recommendation and marketing which are useful for both users and Web site developers. Web log data, which are the record of users' Web-watching behaviors, can be represented as a graph if we regard each visited site as a node and each transition between two sites as an edge. Research for analyzing Web log data has been done by many researchers in the field of Web usage mining: discovering frequent patterns of log data [1][5], modeling users' navigational patterns [2][7], clustering users of specific Web site [4][8], and discovering user communities [3]. As well as visited Web sites, keywords entered to search engines indicate users' interests explicitly. Both

T. Murata and K. Saito: *Extracting Users' Interests of Web-watching Behaviors Based on Site-Keyword Graph,* Studies in Computational Intelligence (SCI) **56**, 139–146 (2007)
www.springerlink.com                    © Springer-Verlag Berlin Heidelberg 2007

entered keywords and visited sites can be obtained from augmented Web audience measurement data (Web log data). A sequence of a user's visited sites and searched keywords can be represented as a directed graph when sites and keywords are regarded as nodes and their adjacency as edges. We call this graph a site-keyword graph in this paper. A site-keyword graph generated from a user's augmented Web audience measurement data is expected to contain his or her interests. In many cases, a user visits several sites and his/her site-keyword graph is really complex. Decomposition of complex site-keyword graph into subgraphs is necessary in order to detect users' interests. This paper describes a method for extracting users' interests by decomposing the above site-keyword graphs. Since users visit their favorite Web sites routinely, dense subgraphs in a site-keyword graph represent their interests. Edges correspond to weak relations are disconnected from a site-keyword graph in order to extract a user's main interest. The qualities of extracted interests are evaluated based on the criteria using PageRank ranking algorithm. Experimental results show that our method succeeds in extracting subgraphs containing many of users' interested sites.

## 2 Graph Representation of Users' Behavior Generated from Web Audience Measurement Data

This section explains Web audience measurement data and augmented ones, which are the record of users' Web-watching behaviors. Site-keyword graphs, graph representation of the above data, are also described.

### 2.1 Web Audience Measurement Data

Web audience measurement data is a kind of client-level Web log data that are recorded at their personal computers. Randomly selected users are asked to use special kind of Web browser that record users' users' Web-watching behaviors (such as visited URL, time of the visit, and elapsed time at the URL) at users' personal computers. An example of Web audience measurement data is shown in Table 1.

Each row of Web audience measurement data represents a user's visit to a URL. Attributes of the data include time, user ID, visited URL, elapsed time, and other miscellaneous information about users' actions. Four companies are doing this sort of investigation in Japan: Nielsen//NetRatings, VideoResearch Netcom, Nikkei BP, and Nihon Research Center. We employ Nielsen//NetRatings raw data (before totaling) since the data contain behaviors of the largest numbers of users among the four Japanese investigation companies. Web audience measurement data are used mainly for statistical analysis, and for detailed investigation of visitors to specific sites. For example, 1) investigation of the users' internet usage (users' age, gender, access time, operating systems of users' PC, and so on), 2) investigation of the relation

**Table 1.** An Example of Web Audience Measurement Data

| time | userID | URL | elapsed time |
|---|---|---|---|
| 00:00 | 9601 | www.jpncm.com/cgi-lib/cmbbs/wforrum.cgi | 10 |
| 00:00 | 9701 | www.dion.ne.jp2 | 7 |
| 00:00 | 3502 | search.auctions.yahoo.co.jp/search | 19 |
| 00:00 | 5201 | eee.eplus.co.jp/shock/shock03.html | 14 |
| 00:01 | 5502 | user.auctions.yahoo.co.jp/jp/show/mystatus | 10 |
| 00:01 | 0501 | user.auctions.yahoo.co.jp/show/mystatus | 6 |
| 00:01 | 3301 | www.pimp-sex.com/amateur/raimi/01/clean.htm | 36 |
| 00:01 | 9701 | auctions.yahoo.co.jp/jp/2⋯.-category-leaf.html | 4 |
| 00:02 | 8501 | www.uicupid.org/chat/csp-room.php | 3 |
| 00:02 | 8001 | page.auctions.yahoo.co.jp/jp/show/qanda | 3 |
| 00:02 | 1501 | www.nn.iij4u.or.jp/ movie/pm/main.html | 11 |
| 00:02 | 9002 | www.umai-mon.com/user/p-category.php | 12 |

between sales promotion campaigns of a company and the number of visitors to the company's Web site, and 3) investigation of the relation between behaviors of buyers at online shops and the results of their offline questionnaire. Users' detailed profiles are given as a table of personal data. Table 2 shows an example of users' personal data. The data contain attributes such as user ID, gender, birth year, birth month, occupation, address, and so on.

**Table 2.** An Example of Users' Personal Data

| UserID | gender | year | month | occupation | address |
|---|---|---|---|---|---|
| 0016 | M | 1971 | 9 | 22 | 3 |
| 0017 | M | 1981 | 9 | 74 | 3 |
| 0019 | M | 1939 | 12 | 94 | 3 |
| 0020 | M | 1950 | 11 | 21 | 3 |
| 0021 | F | 1980 | 3 | 75 | 3 |
| 0022 | F | 1976 | 12 | 95 | 3 |
| 0023 | F | 1975 | 7 | 96 | 3 |
| 0024 | M | 1945 | 5 | 41 | 3 |
| 0025 | M | 1963 | 12 | 13 | 5 |

In general, data for Web usage mining can be divided into two classes: server-level data and client-level data [6]. Web audience measurement data is the latter. As is often pointed out, client-level data reflect users' true behaviors since usages of cached data cannot be recorded on server-level data. Web audience measurement data are therefore suitable for detecting users' true interests.

## 2.2 Augmented Web Audience Measurement Data

Visited URLs of Web audience measurement data are not the only records of a user's Web-watching behaviors. When a user performs search, keywords are given to a search engine. Keywords often play important roles for detecting users' interests as well as visited Web pages. We employ Web audience measurement data augmented with searched keywords for our research. A user's visit to a Web site and an entered keyword to a search engine are both equally recorded as one entry in augmented Web audience measurement data. Searched keywords are not always available because of privacy reasons. Some of the investigation companies do not release keyword data because they sometimes disclose who the users are; users often perform search of their own name and private words. Augmented Web audience measurement data contain such keywords, which are helpful for detecting users' interest. Of course, the authors should pay attention to users' privacies in the process of Web usage mining.

## 2.3 A Site-Keyword Graph

Web audience measurement data can be transformed into a graph structure whose nodes are Web pages and edges are transitions between visits of Web pages. In the same manner, a graph structure of sites and keywords is obtained from augmented Web audience measurement data. We call this graph as site-keyword graph in this paper. In a site-keyword graph, a node represents visited Web site or an entered keyword, and an edge represents transition between a visit to a Web site or an entered search keyword. A site-keyword graph obtained from one user's data represents the users' behavior on the Web. Since a user's favorite sites appear frequently in the augmented Web audience measurement data, nodes representing the favorite sites are densely connected with each other in a site-keyword graph. Searched keywords related to the favorite sites also appear as adjacent nodes in the user's favorite sites. Analyzing a site-keyword graph of a user is expected to clarify the user's interests. Figure 1 shows a part of a site-keyword graph of a user. As mentioned above, each node represents a visited site or an entered keyword. We call the former node as a site node and the latter as a keyword node in this paper. A keyword node is denoted with prefix "Q:", which means a query to a search engine. Site nodes with high degree represent frequently visited sites, and keyword nodes are often connected to the site nodes. If the densely connected site nodes match the user's interests, we can assume that adjacent keywords nodes connected to the site nodes are also closely related to the user's interests. As a way to analyze this site-keyword graph, decomposition of densely connected subgraphs is attempted in this paper.

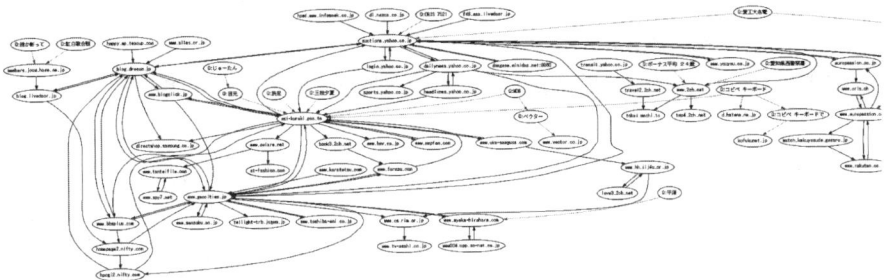

**Fig. 1.** An Example of a Site-Keyword Graph

# 3 A Method for Extracting a User's Interests from a Site-Keyword Graph

In order to extract a user's interests from augmented Web audience measurement data, a method for decomposing a site-keyword graph is proposed in this paper. A user's augmented Web audience measurement data for one month are used for generating a site-keyword graph. Users' interests are extracted from site-keyword graphs by the following four steps. Each step is explained in detail in the following subsections.

1. Applying PageRank algorithm to a site-keyword graph
2. Removing unimportant nodes and edges
3. Decomposing the site-keyword graph by removing edges of weak relations
4. Selecting a subgraph corresponding to a user's main interest

### 3.1 Applying PageRank Algorithm to a Site-Keyword Graph

A site-keyword graph generated from augmented Web audience measurement data is often dense. This is because it represents a user's all transitions on the Web for one month. In order to select users' main interest, PageRank algorithm is applied to a site-keyword graph. PageRank is a well-known ranking algorithm for calculating importance of nodes of a directed graph based on the probabilities of being at each node during random walk.

### 3.2 Removing Unimportant Nodes and Edges

A site-keyword graph contains many noisy nodes and edges that are obstacles for extracting a user's interests. Examples of such nodes are site nodes visited by spywares and site nodes correspond to search engines. Spywares access some specific sites for the purpose of collecting users' information implicitly, which are not intended by users. Nodes corresponds to such sites are removed from a site-keyword graph since they are not related to users' interests.

Site nodes representing search engines (such as Yahoo! and Google) are removed as well because many users visit search engines too often. When a user visits to a search engine, performs a search by entering a keyword, and viewing the searched result, the user visits two pages. This transaction is represented as one keyword node in the corresponding site-keyword graph. The degrees of site nodes of search engines are too big and the site nodes are obstacles for graph decomposition.

As mentioned above, the augmented Web audience measurement data are the data for one month, and they contains many sessions. A session is defined as a sequence of visits on Web page whose elapsed time is less than 30 minutes. Edges connecting sites of different sessions are eliminated as unimportant edges since too much time have been elapsed before and after the transition.

### 3.3 Decomposing the Site-Keyword Graph by Removing Edges of Weak Relations

Edges in a site-keyword graph can be classified to the following four patterns:

1. From a keyword node to a keyword node
2. From a keyword node to a site node
3. From a site node to a keyword node
4. From a site node to a site node

Edges of pattern 1 indicate two continuous keyword searches, and edges of pattern 2 indicate that searched pages are visited. Since these edges represent strong relation between nodes, they are not removed from a site-keyword graph.

Edges of pattern 3 indicate transitions from Web sites to search engines. These edges represent weak relation between connected nodes since the transitions often indicate the beginning of new sequence of Web visits. Edges of this pattern are therefore removed in order to extract subgraphs of related pages.

Edges of pattern 4 indicate transitions between two Web sites. For these edges, the following rule is applied: if more than one transitions between connected two sites are found from augmented Web audience measurement data and elapsed time during the transition is more than zero second, the edge connecting the two sites are not removed, otherwise the edge is removed. Since these edges indicate user's intentional and repeated transitions, they should not be removed.

### 3.4 Selecting a Subgraph Corresponding to a User's Main Interest

A site-keyword graph is decomposed into many subgraphs by the above procedure. A subgraph containing the node of highest PageRank value is selected from the subgraphs as a user's main interest. The subgraph is expected to contain sites and keywords of user's main interest.

# 4 Experiments

Based on the above method, the authors build a system that extract a subgraph of a user's main interest. Used augmented Web audience measurement
data were recorded from December 1 to December 31, 2004 (one month). Total number of users contained in the data is 8155. Maximum number of access
by one user is 2714, while minimum is 1, and average is 134.4. Average size of
extracted subgraphs of our method is 37.4. This size is only 28% of the size of
original site-keyword graphs (37.4/134.4=0.278). As the criteria for evaluating
the qualities of selected subgraphs, nodes whose PageRank values are within
top 10% of all nodes are regarded as "correct answers". The more correct
answers are contained in the selected subgraph, the better the performance is.
Table 3 shows the number of users whose subgraphs are regarded as "good"
(containing more than 50% of correct answers).

**Table 3.** Evaluation of Selected Subgraphs

| # of accesses during one month | all | >100 | >134(average) | >500 | >1000 |
|---|---|---|---|---|---|
| Ratio of selecting good subgraph | 0.27 | 0.42 | 0.48 | 0.73 | 0.90 |

Table 3 shows the percentages that our method succeeds in extracting
users' intentions. The second column (all) shows the percentage of success for
all users, the third column shows the percentage of success for users whose
accesses are more than 100 during one month, and so on. If you choose 28%
nodes randomly from a site-keyword graph, the possibilities of selecting good
subgraph are much lower. Our method succeeds in extracting relatively small
size subgraphs containing many nodes of high PageRank value. The table
shows that the performance is much better for the data of users who access
to many Web sites. For users of more than 1000 accesses during one month,
the percentage of success is 90%. This is because site-keyword graphs for such
users are relatively dense and graph based approach is suitable for such graph.

# 5 Concluding Remarks

This paper describes a method for extracting users' intentions from augmented
Web audience measurement data. Our contributions are 1) Web structure mining method (PageRank) is applicable for evaluating the quality of selected subgraphs and 2) two kinds of data (sites and keywords) can be represented and
processed in the same manner using a site-keyword graph. Although methods for discovering Web communities have been proposed in the field of Web
structure mining, these are not applicable to site-keyword graphs. This is because these methods are based on an assumption that all nodes and edges
in a graph are homogeneous. Components of a graph in real world are often

heterogeneous. A site-keyword graph is composed of two kinds of nodes (site nodes and keyword nodes), and our method removes edges based on simple semantics of connected nodes. Our method is simple and it has abilities of extracting subgraphs related to users' main interests. Our work is expected to give insights for processing noisy heterogeneous graph structures.

# References

1. Berkhin P, Becher J D, Randall D J (2001) Interactive Path Analysis of Web Site Traffic, Proceedings, Seventh International Conference on Knowledge Discovery and Data Mining (KDD01), 414–419
2. Borges J, Levene M (2000) A Fine Grained Heuristic to Capture Web Navigation Patterns, ACM SIGKDD Explorations, Vol.2, No.1, 40–50
3. Murata T (2004) Discovery of User Communities from Web Audience Measurement Data, Proceedings of the 2004 IEEE/WIC/ACM International Conference on Web Intelli-gence (WI2004), 673–676
4. Paliouras G, Papatheodorou C, Karkaletsis V (2000) Spyropoulos, C. D., Clustering the Users of Large Web Sites into Communities, Proceedings of the Seventeenth Interna-tional Conference on Machine Learning, 719–726
5. Pei J, Han J, Mortazavi-asl B, Zhu H (2000) Mining Access Patterns Efficiently from Web Logs, Proceedings of PAKDD Conference, LNAI 1805, 396–407
6. Srivastava J, Cooley R, Deshpande M, Tan P-N (2000) Web Usage Mining: Discovery and Applications of Usage Patterns from Web Data, ACM SIGKDD Explorations, Vol.1, No.2, 12–23
7. Yang Q, Zhang H H, Li T (2001) Mining Web Logs for Prediction Models in WWW Caching and Prefetching, Proceedings, Seventh International Conference on Knowl-edge Discovery and Data Mining (KDD01), 473–478
8. Zeng H-J, Chen Z, Ma W-Y (2002) A Unified Framework for Clustering Heterogeneous Web Objects, Proceedings of the Third International Conference on Web Information Systems Engineering (WISE 2002), 161–172

# Topological aspects of protein networks

J.C. Nacher, M. Hayashida, and T. Akutsu

Bioinformatics Center, Institute for Chemical Research, Kyoto University, Uji, Kyoto, 611-0011, Japan

## 1 Introduction

At the dawn of the present century, the availability of large databases provides a unique opportunity for researchers to investigate the ultimate nature of complex systems. Most often complexity is associated with the presence of large number of components with different types or behaviour, and a huge number of connections between elements. This complex organization is ubiquitous in a large variety of fields spanning from social and technological areas to life sciences and it can be properly studied by means of its network representation. Networks are abstract mathematical objects of real systems composed by nodes connected by edges (or relationships between nodes). Since 1950's random graph theory was extensively used in disciplines such as social sciences and engineering to analyze their data. In random graphs, the degree distribution $P(k)$ (probability that a randomly selected node has exactly $k$ edges) peaks strongly around $K = \langle k \rangle$, where $\langle \rangle$ denotes the average. This distribution is called *poisson* distribution and has guided our thinking in networks for several decades [5]. However, the sunset of twentieth century left us a rich legacy of studies showing that the random graph model could not explain topological properties of real networks. Precisely, two basic features common to many complex systems were found: scale-free topology [2] and small-world structure [19]. In particular, it was shown that the degree distribution of networks composed of experimental data, corresponding to a variety of real systems, followed *scale-free* distribution $P(k) \sim k^{-\gamma}$ with an exponent $\gamma$ between one and four. Social networks, World Wide Web, and biological networks are examples of complex networks that can be characterized by scale-free networks [8, 16, 3, 12, 9, 13, 18, 4], constructed by probabilistically connecting new nodes to highly connected nodes [2]. Other analyses also showed that some networks such as for example, neural network of worm *C.elegans*, film actors and power grid were highly clustered, like regular lattices, yet having small path lengths, like random graphs. The systems with these properties were called "small-world" networks [19].

J. C. Nacher et al.: *Topological aspects of protein networks,* Studies in Computational Intelligence (SCI) **56**, 147–158 (2007)
www.springerlink.com

Over the recent years, major advances in fields of life sciences, together the development of several new technologies, had led to an explosive growth of biological information in both volumen and types of data knowledge at the cellular and subcellular levels. This huge amount of data allows deep and integrative analyses on biological networks and systems, encompassing metabolic, regulatory and signal transduction pathways as well as protein-protein interactions for elucidating functional processes in cells. Cells are structural units that make up higher organisms such as animals and plants, however there are also many single cell organisms. One organism, like the human, stores identical genetic material in every cell. A cell consists of basic chemical elements (Hydrogen, Oxygen, Carbon and others) which interact to form larger molecules (proteins, nucleic acids, carbohydrates and lipids) that make up the basic structural and functional cell of all living matter. Proteins are long chains of aminoacids and perform a wider range of vital functions in cells. Furthermore, they are composed of one or more domains with common properties (see Fig. 1). A protein domain is defined as a building block of the entire protein molecule that is functionally and structurally independent [6]. Concerning protein networks, some works have recently studied the protein interaction networks, revealing the scale-free behaviour of these networks [11, 21]. In addition, a few works carried out some analyses about protein domains ([20, 17]), where the scale-free organization was also found. Moreover, a recent work has shown that domain analysis is helpful to understand and quantify strengths of the protein-protein interactions [10].

In the present work, we first analyze protein similarity networks corresponding to six organisms *Escherichia coli, Saccharomyces cerevisiae, Arabidopsis thaliana, Drosophila melanogaster, Mus musculus* and *Homo sapiens.* In particular, we here present the concept and emergence of *scale-free mixing* which arises in networks where the nodes (proteins) also consist of elemental units (domains). A complete version of this work can be found in [14]. Secondly, by taking advantage of the domain distribution computed previously, we analyze the degree distribution of proteins in networks by attending to functional interactions between proteins. In particular, we investigated the relationship between the emergence of the scale-free topology for the degree distribution and the number of interacting domain pairs. As a main result, we found that a small number of interacting domain pairs play a key role in generating a power-law distribution. In contrast, when a large number of interacting domain pairs is reached, the degree distribution is distorted and the scale-freeness is lost. This finding might be particularly useful for designing artificial networks with scale-free property in biological and non-biological systems. This analysis is being currently extended by developing a mathematical model [15] that reproduces the results shown here in computational experiments.

# Examples of Protein Domains

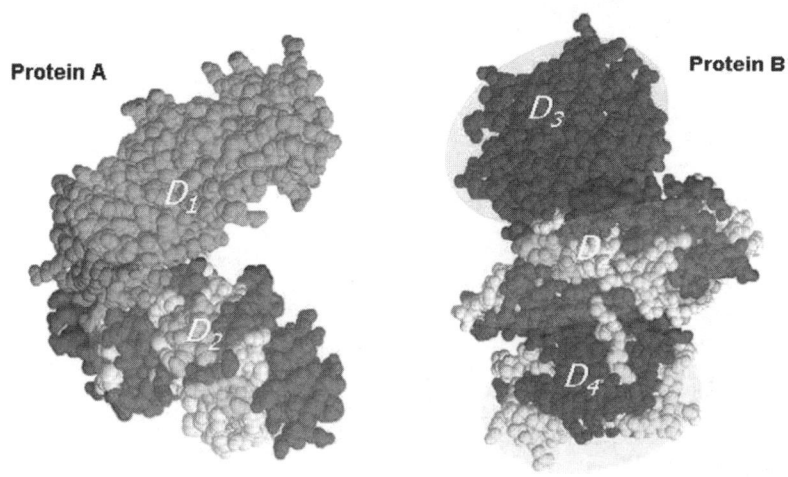

**Fig. 1.** Example of real proteins consisting of several protein domains. Protein A consists of two domains ($D_1$ and $D_2$). Protein B has three domains ($D_3$, $D_2$ and $D_5$). These two proteins have a common domain ($D_2$).

## 2 Scale-free mixing in protein domain networks

We construct an undirected protein domain network corresponding to experimental data from six organisms, where one node of the network corresponds to one protein made of some domains and two proteins are connected by an edge if they share one or more domains. The construction of the network is illustrated in Fig. 2. We used the UNIPROT Knowledgebase database for the individual protein sequences. To be precise, we used the SwissProt part of version 2.5. In this database, the domains of each protein are referred to databases of domains by using their accession numbers. The referred domain databases are *InterPro, Pfam, ProDom, SMART, Prints and ProSite*. Here, we only present the results obtained from *Pfam* databases although similar results were obtained by using the other databases. The spread in node degrees $k$ is investigated by means of the degree distribution $P(k)$, which indicates the number of proteins with specific degree $k$ in the network. Interestingly, the distribution shows not only a power-law with negative exponent $\gamma = -1$, but it resembles the superposition of two power-law functions, one with negative exponent and another with positive exponent $\beta = 1$. We call this distribution pattern *"scale-free mixing"*, which is conserved among six organisms analyzed

# Protein Domain Network

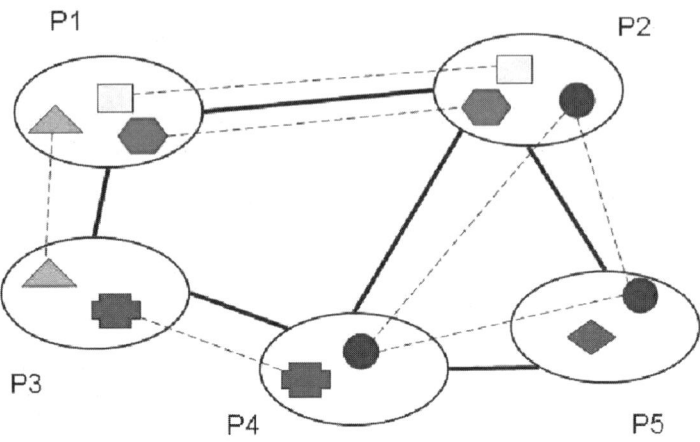

**Fig. 2.** This example illustrates the construction of the protein domain network. We see five proteins, each of which consists of several domains (coloured geometrical symbols). Identical domains are represented by identical symbols. Two proteins are connected by an edge (bold line) if they share one or more domains (dashed line).

from *Escherichia coli* to *Homo sapiens*. We show in Fig. 3 the results corresponding to *M. musculus* and *E. coli* organisms.

## 2.1 Theoretical model

Next, we develop a simple model for explaining the emergence of this superposition of power-laws based on two main ingredients: (1) mutation and (2) duplication of domains. To be precise, the positive power-law branch of the degree distribution can be explained by the emergence in the network of complete subgraphs (i.e., cliques). In particular, these cliques are generated by the duplication mechanism. Therefore, for several values of $k$ a large number of nodes with degree $k$ is generated.

We first generate the domain pattern distribution $P_D(k)$, and by using this distribution we will obtain the degree distribution of the protein domain network $P(k)$ described in Section 2.1.1.

We assume that the initial number of proteins is $N$, each of which consists of only one domain, and being different to each other. Then, the following steps are computed $T$ times: ($i$) A new protein is generated with a new domain with probability $(1 - a)$, ($ii$) Otherwise, one protein is randomly selected and copy

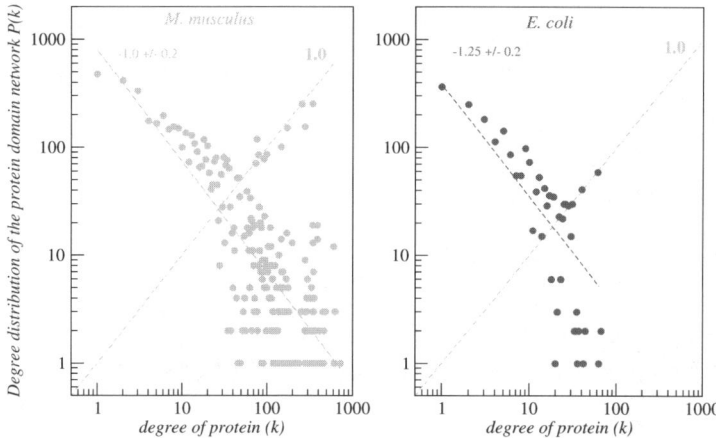

**Fig. 3.** The degree distribution $P(k)$ of the protein domain network in *M. musculus* and *E. coli* organisms displays a similar structure. The distribution resembles a superposition of two power-law functions, one with negative exponent $\gamma = -1$ and another with positive exponent $\beta = 1$ (dashed-lines). Data of protein sequences are from SwissProt, and we used Pfam (circles) protein domain database.

of it is made. In the context of this model, mutation of one protein corresponds to the step $(i)$ and protein duplication is related to the step $(ii)$. There are some points of similarities between this model and the Barabási-Albert (BA) model [2] by considering the following: (1) A node in the BA network model corresponds to a type of domain in our model. (2) The degree of a node in the BA network model corresponds to the number of proteins with same domains in our model. (3) Finally, addition of a new node is related to the creation of a new protein. We denote $i$-th domain as $i$ and the number of proteins consisting of $i$ by $k_i$. Thus, as a copy of a protein is generated with probability $a$ at each time step , and there are $t$ proteins at time step $t$, we can write

$$\frac{dk_i}{dt} = a\frac{k_i}{t} \tag{1}$$

From here, the number of proteins consisting of $i$ reads as:

$$k_i = c\left(\frac{t}{t_i}\right)^a \tag{2}$$

where $t_i$ is the time when the $i$-th domain was first created and $c$ is an appropriate constant. Then, in a similar way than the BA model, we obtain:

$$P_D(k_i = k) \propto k^{[-1-(1/a)]} \tag{3}$$

This equation gives us the domain pattern distribution $P_D(k_i = k)$. To be precise, the distribution indicates the number of different domain pattern.

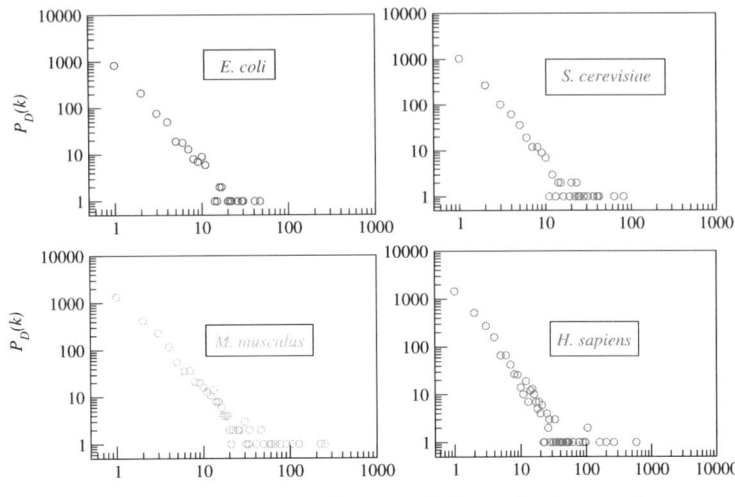

*Number of copies of a protein with the same domain pattern (k)*

**Fig. 4.** The domain pattern distribution $P_D(k)$. Horizontal axis indicates the number of proteins with the same domain pattern. Vertical axis indicates the number of different domain patterns. The results show that a power-law distribution with exponent -2 is conserved in *E. coli*, *S. cerevisiae*, *M. musculus* and *H. sapiens* organisms. Data of protein sequences are from SwissProt, and we used Pfam (circles) protein domain database. Although the results for *A. thaliana* and *D. melanogaster* organisms are not shown, their distributions also follow a power-law with exponent -2.

For each domain $i$, $k_i$ indicates the number of proteins consisting of domain $i$.

Moreover, we can easily change the exponent of the power-law (i.e., $\gamma = [-1 - (1/a)]$) shown in Eq. (3) by modifying the mutation rate $a$. For any mutation rate the value of $\gamma$ is above 2. However, $\gamma$ can be close to 2 if the mutation rate is very small (which is reasonable in biochemical processes) [6]. Finally, we have computed the experimental distribution of domain pattern $P_D(k)$ by using the *UNIPROT-Swissprot* database for protein sequences and *Pfam* for domain databases. Our results show that from *E.coli* to *H.Sapiens* organisms, the distribution follows a power-law with conserved exponent of value 2. In Fig. 4, we show the results for *Escherichia coli*, *Saccharomyces cerevisiae*, *Mus musculus* and *Homo sapiens* organisms. An example of domain pattern distribution is shown in Fig. 5.

## Emergence of scale-free mixing

We here consider a network of proteins, where one node of the network corresponds to one protein and two proteins are connected by an edge if they

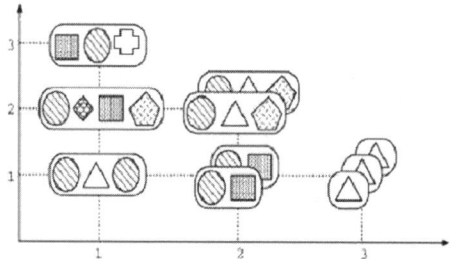

**Fig. 5.** Example of domain pattern distribution $P_D(k)$. Horizontal axis indicates the number of proteins having the same domain pattern (k). Vertical axis indicates the number of different domain patterns $P_D(k)$. In the example, we see one single domain protein (triangle) with three copies (k=3). If only one single domain protein exists, then $P_D(k = 3) = 1$.

share the same domain. Thus, by using Eq. (3) the degree distribution $P(k)$ of this network reads as:

$$P(k) \propto k^{[-(1/a)]} \qquad (4)$$

In Fig. 6, we show the computational results of our model by using $a \sim 0.9$. We can observe a power-law with negative exponent $\gamma = -1$, as it is predicted by Eq. (4). Furthermore, the results of our simulation show a power-law distribution with positive exponent with a value of $\beta = 1$.

We can explain the origin of the power-law branch with positive exponent $\beta = 1$ as follows. One protein can consist of $d$ copies with the same domain after $T$ iterations. Next, we connect two proteins if they consist of the same domain. Then, a cluster of $d$ proteins is generated, where the degree of each protein will be $d-1$. Thus, a power-law branch of the distribution with positive exponent emerges as $P(d - 1) = d$. Interestingly, the signal of this positive power-law distribution was observed for all the analyzed organisms, although is more evident in higher ones as *M. musculus* and *H. sapiens*.

Finally, we also extended our model by adding (3) domain insertion process (i.e., *shuffling domain mechanism*). However, it is worth noting that although the contribution of (3) was relevant for reproducing more precisely the

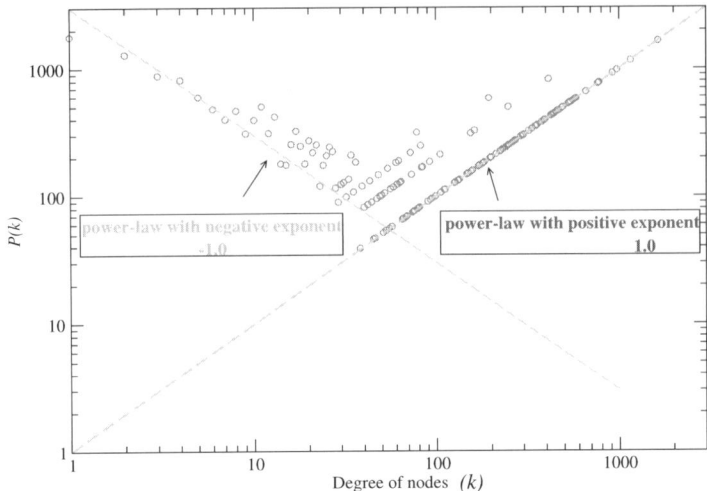

**Fig. 6.** The results of our proposed model by considering only mutation and protein duplication mechanisms. The distribution of nodes with degree $k$ resembles a superposition of two power-law distributions, one with positive exponent $\beta = 1$ and another negative exponent $\gamma = -1$. We call this type of distribution scale-free mixing. $N=100$ initial single domain proteins and up to $T=50000$ iterations.

experimentally observed distribution, in particular for points with high $k$ and low $P(k)$, (1) and (2) mechanisms were enough to re-build (a) the measured scale-free mixing distribution and (b) predict the relevant positive and negative exponents.

## 3 Threshold of interacting pairs for generating power-laws in protein interaction networks

The emergence of scale-free distribution in complex systems is deeply related to the interactions of a large number of elements or agents. In network representation of a given system, where nodes are agents and edges are interactions between agents, the BA model showed that preferential connectivity of a new node to existing nodes and incremental growth of the network are the main causes for the emergence of such power-law distributions. However, we are still far of understanding the origin of preferential attachment in nature. Furthermore, it is also unclear the role of the number of interactions in the construction of scale-free networks and whether the abundance of edges, although uniformly randomly distributed, may change the global architecture of the network. In particular, for example, we may wonder if we can find a threshold for the number of the interacting edges involved in a network beyond which the topology will dramatically change. Here we address this issue by examing the protein interaction networks and present some preliminary results of our computational experiments.

### 3.1 Theoretical approach

We first assume that each protein consists of one domain. For each domain $D$, $n_D$ denotes the number of proteins consisting of domain $D$. From our analysis of protein domain networks, we know that the distribution of domains $P_D(k)$ follows a power-law $k^{-\gamma}$ with exponent $\gamma = 2$ (see Fig. 4). Next, we consider a model for protein interaction which was first introduced by [7] and later successfully applied in [10] for understanding and quantifying strengths of protein-protein interactions. The main idea of this simple model is as follows: If two domains A and B interact, any protein consisting of domain A and any protein consisting of domain B interact. Among all possible domain pairs, we assume that $N$ domain pairs are selected as interacting domain pairs with uniformly random distribution.

### 3.2 Preliminary computational results

We performed computational experiments as follows. We generated domains according to the power-law $k^{-\gamma}$ with exponent $\gamma = 2$, where the abundance of proteins consisting of the same domain are gathered in an integer array. In total, 163444 different families of domains were created for $k = 1,...100$, where $k$ indicates the number of proteins with the same domain pattern. We then randomly select two domains and connected the corresponding protein pairs with an edge. This process was repeated $N$ times. The results are shown in Fig. 7. We can see that the degree distribution follows a power-law if the number of interacting domain pairs $N$ is small. However, the computational experiments also showed us that the value $N$ contains a threshold beyond which the topology dramatically change. As we can see in Fig. 7(b-c), the power-law degree distribution first is distorted and, later, when a larger value of $N$ is reached, the scale-freeness is completely vanished.

We have used the DIP database for constructing the protein interaction network of several organisms. We here report the results for *Saccharomyces cerevisiae, Drosophila melanogaster* organisms. In both cases, we found that a power-law distribution was found and the exponent is close to the value 2, the same that we obtained in our computational experiment shown in Fig. 7. We are currently extending the present analysis and developing a complete mathematical model [15] that explains the threshold point between power-law and distorted distributions in protein interaction networks.

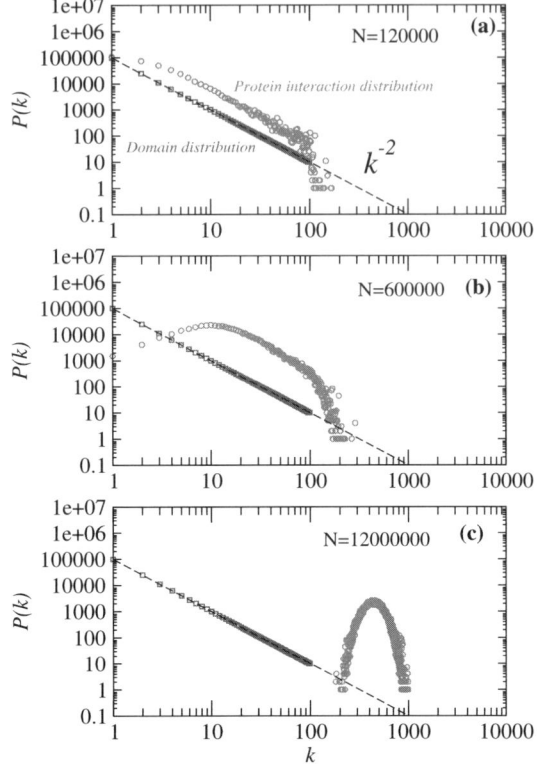

**Fig. 7.** Degree distribution for protein domain (squares) and protein interaction networks (circles) for three values of interacting domain pairs $N$ shown in figures. Dashed line corresponds to the theoretical power-law $k^{-\gamma}$ with exponent $\gamma = 2$.

## 4 Summary

Our analysis on domain similarity networks has shown that the distribution exhibits not only a power-law with negative exponent $\gamma = -1$, but it resembles the superposition of two power-laws, one with negative exponent and another with positive exponent $\beta = 1$. Moreover, the results revealed that the emergence of this superposition of power-laws requires only at least two main ingredients: (1) mutations and (2) genetic duplication of domains, which reflects the importance of these mechanisms occurred during the evolution.

We should also bear in mind the striking difference between the emergence of distribution branches with positive and negative exponents. In particular, the negative branch appears as a result of statistical abundance of nodes with a large number of edges (i.e., hubs). In contrast, as we have illustrated with our model, the positive branch of the distribution emerges due to the existence of complete subgraphs.

As future research work, it would be interesting to investigate whether this *scale-free mixing* pattern is only a feature of the protein domain network or it could also appear in different biological systems or artificial and technological networks (for example, communication networks such as domain servers and Internet).

On the other hand, the existence of a threshold of interacting domain pairs for generating power-laws in protein interaction networks might be an important property to take into account for developing and designing artificial networks for technological and communication systems with scale-free property. Surprisingly, the deeper our understanding goes with regard to genomics and subcellular biological networks, more striking similarities are found with respect to technical networks. However, there is still room for progressing and improving the efficiency of the artificial networks by exporting properties that are only present in biological networks. For example, metabolic, protein interaction and signalling architectures, adapted during the evolution to survive rapid and drastic changes in its enviroment, may provide a fruitful source of inspiration for dealing with important design aspects of networks such as robust and error/attack tolerance of scale-free network. It is known that although scale-free networks are fragile to a purposeful attack of highly connected nodes, networks with scale-free property are robust against random node failures [1]. Thus, the study of biological networks may not only feed a deeper knowledge on molecular biology, but may also provide a rich source of ideas on large-scale artificial communication networks.

**Acknowledgement:** This work was partially supported by Grant-in-Aid "Systems Genomics" from MEXT (Japan) and by the Japan Society for the Promotion of Science (JSPS)

# References

1. Albert R, Jeong H and Barabási A-L (2000) Error and attack tolerance of complex networks Nature **406**:378–382.
2. Barabási A-L and Albert R (1999) Emergence of scaling in random networks. Science **286**:509–512.
3. Barrat A, Barthelemy M, Pastor-Satorras R and Vespignani A (2004) The architecture of complex weigthed networks. Proc. Natl. Acad. Sci. USA **101**(11):3747–3752.
4. Barabási A-L and Oltvai Z N (2004) Network biology: Understanding the cell's functional organization. Nature Reviews Genetics **5**:101–113.
5. Erdös P and Rényi A (1960) Random Graphs. Publ. Math. Inst. Hung. Acad. Sci. **5**:17–63.
6. Doolittle R F (1995) The multiplicity of domains in proteins. Ann. Rev. Biochem. **64**:287–314.

7.  Deng M, Mehta S, Sun F and Chen T (2002) Inferring domain-domain inter-
    actions from protein-protein interactions. Genome Research **12**:1540–1548.
8.  Dorogovtsev S N and Mendes J F F (2003) Evolution of Networks: From Bio-
    logical Nets to the Internet and WWW, Oxford University Press, Oxford.
9.  Furusawa C and Kaneko K (2003) Zipf's law in gene expression. Phys. Rev.
    Lett. **90**:008102(1–4).
10. Hayashida M, Ueda N and Akutsu T (2003) Inferring strengths of protein-
    protein interactions from experimental data using linear programming. Bioin-
    formatics **19**(2):58–65.
11. Jeong H, Mason S, Barabási A-L and Oltvai Z N (2001) Lethality and centrality
    in protein networks. Nature **411**:41–42.
12. Jeong H, Tombor B, Albert R, Oltvai Z N and Barabási A-L (2000) The large-
    scale organization of metabolic networks. Nature **407**:651–654.
13. Kuznetsov V A, Knott G D and Bonner R F (2002) General statistics of stochas-
    tic process of gene expression in eukaryotic cells. Genetics **161**:1321–1332.
14. Nacher J C, Hayashida M and Akutsu T (2006) Physica A, in press.
15. Nacher J C, Hayashida M and Akutsu T *In preparation*.
16. Pastor-Satorras R and Vespignani A (2004) Evolution and Structure of the
    Internet: A Statistical Physics Approach, Cambridge University Press, Cam-
    bridge.
17. Qian J, Luscombe N M and Gerstein M (2001) Protein-fold occurrence in
    genomes: power-law behaviour and evolutionary model. Journal of Molecular
    Biology **313**:673–681.
18. Ueda H R, *et al.* (2004) Universality and flexibility in gene expression from
    bacteria to human. Proc. Natl. Acad. Sci. USA **101**(11):3765–3769.
19. Watts D J and Strogatz S H (1998) Collective dynamics of "small-world" net-
    works. Nature **393**:440–442.
20. Wuchty S (2001) Scale-free behavior in protein domain networks. Mol. Biol.
    Evol. **18**(9):1694–1702.
21. Yook S H, Oltvai Z N and Barabási A-L (2004) Functional and topological
    characterization of protein interaction networks. Proteomics **4**:928–942.

# Collective Intelligence of Networked Agents

Akira Namatame

Department of Computer Science, National Defense Academy,
Yokosuka, Kanagawa, 239-8686, Japan
nama@nda.ac.jp, http://www.nda.ac.jp/~nama/

**Abstract:** Social systems have emergent properties that cannot be trivially derived from the properties of their members. Very simple rules concerning individual behavior can result in very unpredictable outcomes on the collective level. Emergent phenomena at the aggregate level influence in turn individuals and their interactions. This bi-directional causal relationship is at the essence of understanding complexity of the social systems. In this paper, we study the issue of emergent collective intelligence. We model networked agents who repeatedly play the games with three strategies. We show that networked agents collectively evolve efficient rules that realize a Pareto-efficient outcome of the underlying games. The essential point is that desirable behavioral rules are emerged spontaneously at the collective level from the pair-wise interactions of networked agents.

## 1 Introduction

Why the colony of ants works collectively as it does, and an effectively as it does? Collective efforts are a trademark of both insect and human societies. They are achieved through relatedness in the former and some unknown mechanisms in the latter. How we can extend the concept of collective intelligence observed in social insects is to human societies [1]. The problem of achieving cooperation among human has been described as social dilemma, prophesying the inescapable collapse of many human enterprises. An important key to answer this question is to look at pair-wise interactions among human. The distinction between anonymous and individualized societies provides a base for distinguishing between societies of ants and humans. Individuals have different properties perceived by others. They occupy different social roles and social positions [5]. Human societies can be characterized in terms of power and trust relations. The structure of human societies consists both of groups such as local communities and of social networks that link individuals,

A. Namatame: *Collective Intelligence of Networked Agents,* Studies in Computational Intelligence
(SCI) **56,** 159–176 (2007)
www.springerlink.com

groups and organizations. Individual properties by which individuals differ provide the basis on which social structure is built. Even meaningless features acquire meaning in social interactions if we are perceived by others [16].

For the last decade, attempt been made to develop some general understanding, and ultimately the theory of collective systems that consist of a large collection of agents. It is common to call desirable emergent properties of collective systems, *"collective intelligence"* [22]. Interactions are able to produce collective intelligence at the macroscopic level that is simply not present when the components are considered individually.

In his book, titled The Wisdom Of Crowds, Surowiecki explores a simple idea that has profound implications: a large collection of people are smarter than an elite few at solving problems, fostering innovation, coming to wise decisions, and predicting the future [20]. His counterintuitive notion, rather than crowd psychology as traditionally understood, provides us new insights for understanding how our social and economic activities should be organized. He explains the wisdom of crowds emerges only under the right conditions: (1) diversity, (2) independence, (3) decentralization, and (4) aggregation. These conditions are also keywords for the community of the agent research.

However, the fact that selfish behavior may not achieve full efficiency has been well known in the literature. Recent research efforts have focused on quantifying the loss of system performance due to selfish and uncoordinated behavior. The degree of efficiency loss is known as the price of anarchy [21]. Many social systems are based on an analogous assumption that individuals are selfish optimizers, and we need methodologies so that selfish-behaviors of individuals need not degrade the system performance. Of particular interests is the issue how social interactions should be restructured so that agents are free to choose their own actions while avoiding outcomes that none would choose.

Research in social sciences suggests that interaction between individuals, does not just involve sharing information but, in large part, its function is to construct a shared reality that consists not only of shared information but also of agreed upon opinions. In this process, individuals do not simply transmit information, but more importantly, they influence one another to arrive at a common interpretation of information. In contrast with a simple spread of information hearing the same opinion from a number of sources results in a higher probability of adopting this opinion, than hearing it form a single source. It also matters whom individuals hear this information form. Social influence is usually defined as a change in individual's thoughts feelings or actions resulting from the real or imagined presence of others. Social influence concerns not only formation of opinions but also a variety of other social phenomena such as learning from observation i.e. social modeling, attitude changes and norm formation [23].

An evolutionary approach is needed for designing social systems so that selfish individual behaviors need not degrade the system performance by collectively evolving social norms. Darwinian dynamics based on mutation and selection form the core of models for evolution in nature [10][17]. Evolution through natural selection is understood to imply improvement and progress. If multiple populations of agents are adapting each other, the result is a co-evolutionary process. The problem to contend with in co-evolution based on the Darwinian paradigm, however, is the possibility of an escalating arms race with no end. Competing agents might continually adapt to each other in more and more specialized ways, never stabilizing at a desirable outcome [15].

In this paper, I propose the concept of collective evolution in a society of networked agents. In biology a better gene individual is the unit of selection. However, collective evolutionary process is expected to compel agents towards ever more refined adaptation and evolution resulting in sophisticated behavioral rules rather than strategies. The persistence and sustainability of a society of the networked agents in turn depends on its persistent collective evolution. Therefore, the mission of collective evolution is to harness the complex systems of selfish agents and to serve to secure a sustainable relationship in an attainable manner so that desirable properties could emerge as collective intelligence. Of particular interests is the question how social interactions can be restructured so that agents are free to choose their own actions while avoiding outcomes that none would have chosen. It is shown that a collection of interacting agents evolves into both efficient and equitable outcome.

An interesting problem, which has been widely investigated, is under what circumstances will agents by learning converge to some particular equilibrium? [15]. We want to endow our agents with some simple way of behavioral rules and describe the co-evolutionary dynamics that magnifies tendencies toward better outcomes. By incorporating a consideration of how agents interact into models we not only make them more realistic but we also enrich the types of aggregate behavior that can emerge. The framework of collective evolution distinguishes from co-evolution in three aspects. First, there is the coupling rule: a deterministic process that links past outcomes with future behavior. The second aspect, which is distinguished from individual learning, is that agents may wish to optimize the outcome of the joint actions. The third aspect is to describe how a coupling rule should be improved with the criterion of performance to evaluate how the rule is doing well.

The social space consists of networks of self-interested agents, continuous evaluations of their performance as well as their behavioral rules. Behavioral rules are here treated as the constraints on individual action and they specify the action choice based on the specific outcomes. The learning of new behavioral rule, and the strife of each agent to act in keeping with the coupling with the Social norms are here treated as the shared behavioral rules that constitute common constraints on all individuals in a society. For agents in a social context to achieve collective intelligence, it is a continuous process that requires social behavior based on social rationality [13]. To in turn achieve social rationality requires for individually rational behavior to be constrained by some obligations. Instead, the analysis of what is chosen at any specific time is based upon an implementation of the idea that effective behavioral rules are more likely to be retained than ineffective ones.

## 2 Game Theoretic Models of Social Interactions

Social interdependence defines the type of interaction between individuals in a society. Social interdependence can be understood as a dependence of outcomes of one individual on another individual behavior. Such a relationship between payoffs for choices of different individuals is usually described with the formalism of the game theory. In this formalism choices of one agent change the payoff structure of other agents.

Prisoner's dilemma is a mathematical model of conflict between individual and group interests. In prisoner's dilemma each individual can either compete or collaborate with a partner. The payoff for each individual is always higher if the individual chooses to compete. On the group level, however, the payoff is always higher if the individuals cooperate. In multi-agent models agent's decisions are dictated by agent's strategies. Since on the individual level competition is the dominating strategy, the main question is how can cooperation between individuals emerge? To answer this question one can equip the agents with pre-specified strategies, behavior rules and properties and observe to consequences of their interactions as they play the game repeatedly, or allow for the evolution of strategies by genetic algorithms. The famous result obtained by Axelrod is that cooperation will eventually emerge even in a society of egoistic individuals, as these pairs of players that will be able to converge on cooperation will gain more that those that will be locked in competition [2]. The introduction of genetic algorithms enabled researchers to investigate the natural selection of social behaviour using sophisticated computer simulations.

Simulation results concerning conditions that facilitate the emergence of cooperation have important practical implications. Research has focused almost exclusively on the Prisoner's Dilemma game, a type of social interaction between two individuals in which both players benefit if both cooperate, but each can do even better, at the expense of the other, by defecting. prisoner's dilemma occur ubiquitously in naturally occurring human and animal interactions, and the players often cooperate. Since interdependence defines one of the most fundamental ways to link individuals many different issues have been addressed in the paradigm of prisoner's dilemmas. The results concern not only strategies, individuals features and interaction patterns that facilitate vs. inhibit cooperation in the group, but also the formation of social structures, like the emergence of solidarity and mutual help networks, power structures, patterns of settlements, trust networks etc.

### 2.1 A Basic Dilemma Game

Consider a population of $N$ agents, each faces a binary choice problem between two behavioral types: Cooperate (C) or Defect (D). For any agent the payoff to a choice of C or D depends on how other agents also choose C or D. The payoffs to each agent choosing from Cooperate (C) or Defect (D) are given in Table 1.

In the prisoner's dilemma, two players have the choice to cooperate or to defect. Both obtain payoff $R$ *(reward)* for mutual cooperation, but a lower payoff

*P(punishment)* for mutual defection. If one individual defects, while the other cooperates, then the defector receives the highest payoff T (temptation) whereas the cooperator receives the lowest payoff S (sanction). Many works on evolution of cooperation have been focused on two-person dilemma games with the payoff parameters satisfying: (1) T > R > P > S, (2) 2R> S + T [2][10]. The inequality (1) is the condition for that if each agent seeks his individual rationality; they result in choosing (D, D). The outcome is better for each if both cooperate than if both defect, but each agent gets the best payoff by defecting while the other cooperates. In that case, the cooperative bird gets the worst payoff, expending time and energy but getting nothing in return.

**Table 1.** The payoff matrix of a dilemma game

| Agent A \ Agent B | Cooperate(C) | Defect (D) |
|---|---|---|
| Cooperate (C) | R | S |
| Defect (D) | T | P |

To cooperate successfully, the players have to sacrifices their individual rationality. The main obstacle for the evolution of cooperation is that natural selection favors defection in most settings. Understanding the mechanisms that can lead to the evolution of cooperation through natural selection is a core problem in social sciences and biology. The evolutionary problem is to explain how such social behaviour could have evolved, given that natural selection operates at the level of the individual organism or gene. This problem has been largely solved by the theories of reciprocal altruism and indirect reciprocity, and computer simulations have shown that, after thousands of repetitions, reciprocal strategies such as Tit for Tat (cooperate if and only if the other player cooperated last time) or Pavlov (repeat any action that led to a good outcome last time, otherwise switch) tend to evolve, resulting in widespread joint cooperation.

Here, we study models of cooperation that are based on two extended dilemma games with three strategies.

## 2.2 Public Good Game with Voluntary Participation

The problem of achieving cooperation among self-interested peoples has been also described as the public good game. It has been implicitly assumed compulsory participation in the dilemma good games. In many situations, however, individuals often may drop out of unpromising and risky situations and instead rely on the perhaps smaller but at least secure earnings based on their individual efforts. Hauert et al show the importance of volunteering in public good games [9]. Such optional participation is implemented by considering a third strategic type, the loners. Loners are risk averse and do not engage in the prisoner's dilemma game but rather rely on small but fixed earnings. With including optional participation, the original dilemma game is formulated as the game with three strategies as shown in the payoff matrix in Table 2. In this case, rock-paper-scissors-type cyclic dominance of strategies can arise whenever the participation in a dilemma game is optional, since cooperator beats loner, loners beats defector, and defector beats cooperator [11].

In their work, evolutionary game theory is designed to capture the essentials of the population. With optional participation in the public goods game, loners who do not join the group, defectors, cooperators and loners will coexist through rock-paper-scissors dynamics. If defectors have the highest frequency, loners soon become most frequent, as do cooperators after loners and defectors after cooperators. The cyclic dominance of the three strategies is reflected in the cycle, a closed trajectory that contains fixed points. They claim that this simple but effective mechanism does not require sophisticated strategic behavior at individual levels. Voluntary participation can foil exploiters and overcome the social dilemma.

**Table 2.** The payoff matrix with loners

| Agent B / Agent A | $S_1$ (Cooperator) | $S_2$ (Defectors) | $S_3$ (Loners) |
|---|---|---|---|
| $S_1$ (Cooperators) | R          R | T          S | σ          σ |
| $S_2$ (Defectors) | S          T | P          P | σ          σ |
| $S_3$ (Loners) | σ          σ | σ          σ | σ          σ |

## 2.3 Dilemma Game with Meta-Strategy

In the repeated prisoner's dilemma, two individuals interact several times, and, in each round, they have a choice between cooperation and defection. As is standard in repeated games, new strategies become possible when the game is repeated, and these strategies can lead to a wider range of equilibrium outcomes. In particular, in the infinitely repeated prisoner's dilemma, cooperation becomes an equilibrium outcome, but defection remains an equilibrium as well. Ever since Axelrod's computer tournaments [2], TFT is a world champion in the repeated prisoner's dilemma, although it has some weaknesses and has at times been defeated by other strategies.

Imhof et al analyzes the evolutionary dynamics of three meta-strategies for the repeated prisoner's dilemma: always defect (ALL-D), always cooperate (ALL-C), and tit-for-tat (TFT) [11]. TFT cooperates in the first move and then does whatever the opponent did in the previous move. TFT is a conditional strategy, whereas the other two strategies are unconditional. Therefore, it is natural to include a complexity cost for TFT. Therefore they reduce the payoff for TFT by a small constant, and consider the payoff matrix with three meta-strategies in Table 3. From the payoff matrix in Table 3, ALL-D is the only evolutionarily stable strategy (ESS) and the only strict Nash equilibrium. If everybody uses ALL-D, then every other strategy has a lower fitness. Therefore, no mutant strategy can invade an ALL-D population. In contrast, neither TFT nor ALL-C nor any mixed population has this property.

However, the pair-wise comparison of the three strategies leads to the followings. (1) ALL-C is dominated by ALL-D, which means it is best to play ALL-D against both ALL-C and ALL-D. (2) TFT is dominated by ALL-C. These two strategies cooperate in every single round, but the complexity cost of TFT implies that ALL-C has a higher payoff. (3) If the average number of rounds exceeds a minimum value, then TFT and ALL-D are bi-stable. This result means that, choosing between ALL-D and TFT, each meta-strategy is a best response to itself. Despite ALL-D being the only strict Nash equilibrium, they observe evolutionary oscillations among all three meta-strategies like a Rock-Paper-Scissors-type cyclic game. The population cycles from ALL-D to TFT, and TFT to ALL-C and back to ALL-D. The time average of these oscillations can be entirely concentrated on TFT, and therefore they conclude that favor cooperation over defection.

**Table 3.** The payoff matrix with meta-strategies

| Agent B / Agent A | $S_1$ (ALL-C) | | $S_2$ (ALL-D) | | $S_3$ (TFT) | |
|---|---|---|---|---|---|---|
| $S_1$ (ALL-C) | | R | | T | | R |
| | R | | S | | R | |
| $S_2$ (ALL-D) | | S | | P | | P |
| | T | | P | | P | |
| $S_3$ (TFT) | | R | | P | | R |
| | R | | P | | R | |

# 3 Strategy Choice Based on Learnable Behavioral Rule

In orthodox rational choice theory, agents are modeled as cognitively sophisticated and entirely self-interested decision makers who evaluate every future consequence of possible actions and select the action alternative that maximizes own payoffs. Discrete choice analysis grounded in the theory of utility maximization has proven quite successful in terms of its usefulness. However, this approach is being challenged by a line of research originating in cognitive psychology that is causing economists to re-examine the standard model of choice behavior. In the words of the psychologist Kahneman, economists have preferences; psychologists have attitudes [23].

However, experimental evidence supports the view the behavioral rules are the proximate drivers of most human behavior. The rule-governed action can be also pictured as a quasi-legal process of constructing a satisfying interpretation of the choice situation. The behavioral rules we do use are essentially defensive ones, protecting us from mistakes that perceptual illusions may induce. However, the question remains as to whether behavioral rules themselves develop in patterns that are broadly consistent with the rational model postulates. This is a vital scientific concern. If there are preferences behind the formation of behavioral rules, then how they are correlated with these underlying preferences.

We seldom do new things. Most behaviors are repeated, but many researchers do not pay much attention to this aspect. Few would dispute the claim that most behaviors are repetitive, yet in spite of a large literature on learning, the habit concept has received only minor attention. The sort of coordination problems we have in mind are those that we commonly solve without thought or discussion-usually so smoothly and effortlessly that we don't even notice that there is a coordination problem to be solved.

Verplanken and Aarts define habits as learned sequences of acts that have become automatic responses to specific cues, and are functional in obtaining certain goals or end-states [23]. Obviously, many behaviors may fall under this definition, varying from being very simple to being complex. Habits are learned sequences of actions. Habits are also automatic responses to specific cues. Habitual acts are also instigated as immediate responses to specific situations. These responses occur without purposeful thinking or reflection and often without any sense of awareness. Most habits are created and maintained under the influence of learning. For instance, behavioral rule that has positive consequences is more likely to be repeated, whereas negative consequences make repetition less likely. Repeated behaviors may turn into habits, which are automatic responses to specific cues and are functional in obtaining certain goals. We may want new and desired behaviors to become habits, which makes them stable and difficult to change. Habituation thus becomes a behavioral rule.

Social norms and habit influence in turn individuals' purposive behaviors based on their current preferences. This bi-directional causal relationship is at the essence of the study of the cognitive decision-making process. Understanding the nature of the relationship between two different levels at which actual choice is a grand challenge of cognitive science. Explanation of this relationship calls for examining the types of social interactions that link individuals in social contexts.

In this paper we propose a hybrid choice model based on both rule-based choice and preference-based choice. Agents adhere to behavior rules via local adaptation of behavior. The adaptation of behavior rules consists of an internalization of social norms, or more precisely a synchronization of the individual behavioral rule to those of the other neighbors. Each agent applies the hybrid choice model based on both agent specific assessments of the situations (rational choice model) and social norms or habits (rule based choice model). Social norms have been treated here as constraints on agent-specific rational choices. Each agent is modeled to evolve her behavioral rule. This hybrid choice model at individual levels is the core of emergent socially intelligent behavior.

We stresses that the performance of the socio-economic system consisting of self-fish agents depends on how they are properly coupled. A strategy choice based on the behavioral rule for repeated play of the game uses the recent history of play to choose one of the three strategies for the next play. Here, we assume that each agent can refer to the last outcome. Each behavioral rule is represented as a binary string so that the genetic operators can be applied. We represent a behavioral rule by a 3-bits string using $S_1=0$, $S_2=1$ and $S_3=2$. In order to accomplish this we use a bit string. Since no memory exists at the start of the game, extra one bit is needed to specify a hypothetical history at the beginning.

Each strategy was essentially a Moore machine designed to generate a move on the basis of any possible three-move history. For each outcome, a player receives one of nine payoffs in Table 2 and Table 3, thus there are 9 different one-move histories. A string of 9 binary digits therefore suffices to specify a choice for every one-move history. A hypothetical one-move history was also necessary to get each game started. For this, 3 additional binary digits, named the premise genes, were added, making a total genome of 12 binary digits.

The offspring strategies that played in each subsequent generation were formed from the most successful strategies of the previous generation, using a genetic algorithm.

Each position $p_j$, $j=1,...,12$, in Figure 2 represents as follows. The first position $p_1$ encodes the initial strategy that the agent takes at each generation. A position $p_j$, $j=2,3$, encodes the history of mutual hands (rock, scissors, or paper) that agent and her opponent took at the previous round. A position $p_j$, $j=4,..,14$, encodes the action that the agent takes corresponding to the values at the positions $p_j$, $j=2,3$. Since there are nine possible outcomes for each round. We can fully describe a behavioral rule by recording what the strategy will do in each of the nine different outcomes that arise in the last play of the game. A rule must specify depending each outcome, what strategy the agent should choose at the net round. Since there are three strategies, the number of possible behavioral rules is $3^9$. The hope is that agents would find a better behavioral rule out of the overwhelming possible rules after a reasonable number of plays.

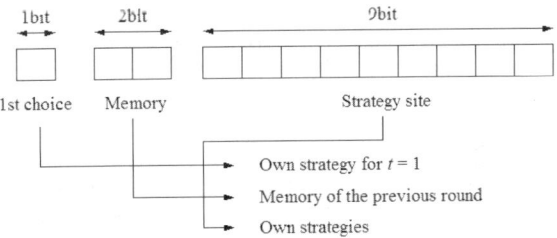

**Fig. 1.** An agent's behavioral rule with a memory of one round

# 4 Individual Learning vs. Social Learning

To achieve desirable outcomes, a primary question is how each individual should learn in the context of many learners [25]. There are two competing approaches for describing the learning model of the population: the microscopic model based on individual learning and the macroscopic model based on social learning.

In the category of individual learning, agents are modeled to have some repertories of behavioral rules, and they update those rules using the existing rules within as shown in Figure 2(a). Natural selection operates on the local probability distribution of behavioral rules within the repertoire of each individual agent. In an individual learning model, we could say that each agent checks if another randomly chosen agent in the population gets a higher payoff, and, if so, switches to that behavior with a probability proportional to the payoff difference.

There is no imitation or exchange their experiences among agents in individual learning. On the other hand, social learning becomes valuable in a social context, since it can help to surface new ideas and generate social consensus on issues that no single individual can effectively make right decision about alone. Social learning can also be extended beyond the boundaries of a single agent. Social learning is one that has an internal process for cultivating individual learning and connecting it to others. So when faced with change, a collective has the requisite energy and flexibility to move in the direction it desires.

In an orthodox social learning model as shown in Figure 2(b), agents play based on the prescribed behavioral rules. The summed payoff of each game provides the agent's fitness. After every individual has played the game with her neighbors, each rule of the agents is updated according to the general evolutionary rules, and the behavioral rule is crossover with the most successful behavioral rule of her neighbors. Their success depends in large part on how well they learn from their neighbors. If an agent gains more payoff than her neighbor, there is a chance her behavioral rule will be imitated by others.

The principle of social learning itself can be thought of as the consequence of any one of three different mechanisms. It could be that the more effective individuals are more likely to survive and reproduce. A second interpretation is that agents learn by trial and error, keeping effective rules and altering ones that turn out poorly. A third interpretation is that agents observe each other, and those with poor performance tend to imitate the rules of those they see doing better.

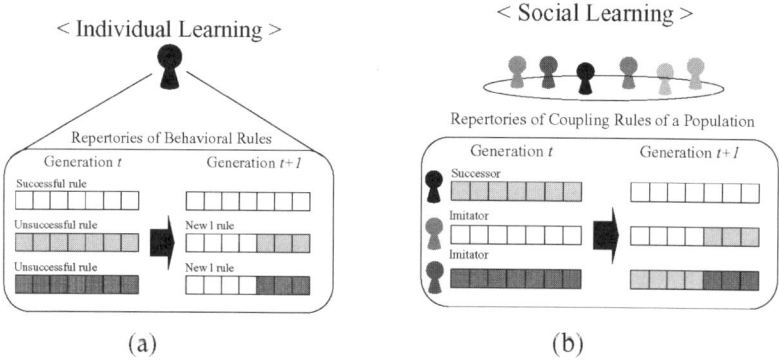

**Fig. 2.** Individual learning vs. social learning with evolvable behavioral rules

The most unrealistic aspect of the rule learning is the large number of strategies each agent considers. Even if the set of rules is limited to very simple ones, each agent remembers to many strategies. A realistic model should account for the fact that agents consider a much smaller number of rules from which they learn and make decisions; and that the rules agents consider are often preconditioned by factors such as imitation that have evolved over the generations.

# 5 Social Interactions of Networked Agents

For the emergence of group level phenomena the pattern of social interactions is critical global interactions. The interaction structure specifies who affects whom, and this network structure may vary from one individual to another. Social inter-dependence can be understood as a dependence of outcomes of one individual on another individual behavior. Such a relationship between payoffs for choices of different individuals is usually described with the formalism of the game theory.

A crucial ingredient in social interaction models is the network structure in which individuals interact. The interaction structure specifies who affects whom, and this network structure may vary from one individual to another. The agents involved would learn two things: with whom to interact and how to behave. That is to say that learning dynamics operates both on network structure and strategy. The interaction structure specifies who affects whom.

In order to describe the ways of interaction, the random matching model is frequently used. In the random matching model, in which each agent is assumed to interact with a randomly chosen agent from the population. There are also a variety of interaction models, depending on how agents meet, and what information is revealed before interaction. Some social processes and some simulation models assume global pattern of interaction. Everyone interacts with a certain probability with everyone else in the social group.

Individuals in most social processes do not interact with everyone with equal probability. Space imposes important constrains on interaction. Individuals interact with the highest probability with others who are nearby. Empirical research has shown, for example, that the probability of interactions drops down as a square of the distance between the places where each individual lives. There are many situations in which a spatial environment becomes a more realistic representation, since interactions in real life rarely happen on such a macro-scale as assumed in the global interaction model. Spatial interaction is generally modeled through the use of the two dimensional (2D) grid in Figure 3(a) with each agent inhabiting each cell of the lattice on the grid. Interaction between agents is restricted to nearest neigh-boring agents. Each agent chooses an optimal strategy based on local infor-mation about what her neighbors will choose. However, the consequences of their choices may take some time to have an effect on agents with whom they are not directly linked. Each agent at each cell in this figure is characterized by its location and its state. In cellular automaton the interaction rules are local, in that the state of each cell depends on the state of neighboring states in a way specified by a specific rule. Cellular automata are most useful for discovering spatial and spatio-temporal patterns that emerge from simple rules of locally defined interactions. And even very simple rules can produce amazingly complex dynamics, and no direct relationship exists between the complexity of the rules and the complexity of the resultant dynamics.

Social interactions are governed by formal and informal networks. The approach of social networks formalizes the description of networks of relations in a society. In social networks individuals or at a different level of description social groups can be represented as nodes, and relations or flows between individuals or groups can be portrayed as links. These links can represent different types of relations between individuals such as friendship, trust, exchange of information, and transfers of knowledge.

At another end of the spectrum we have models where individuals interact with both fixed their neighbors and randomly chosen agents from the population. Watts and Storogatz introduced a small-world network architecture that transforms from a coupled system with nearest neighbors to a randomly coupled network by rewiring the links between the nodes [24]. For instance, consider a two-lattice model in which each node is coupled with its nearest neighbors, as shown in Figure 3(b)(c). If one rewires the links between the nodes with a small probability, then the local structure of the network remains nearly intact.

If we fix the interaction structure, we get models of the evolution of strategies in games played on a fixed network structure. Ultimate interest resides in the general case where structure and behavioral co-evolve and ask the basic question: whether co-evolution of structure and behavioral rules supports or reverses the conventional wisdom about equilibrium selection in the repeated games.

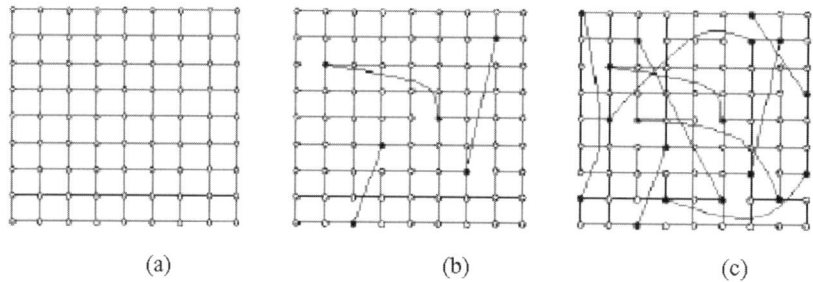

(a)                          (b)                          (c)

**Fig. 3.** Social interactions of networked agents. (a) p = 0, a regular lattice. (b) p = 0.1, some of the links have been re-wired resulting in a small–world network. (c) p = 0.5, additional re-wiring has occurred. As p approaches 1, a transition to a random network will occur

# 6  Simulation Results

## 6.1 Public Good Game with Voluntary Participation

Figure 4 shows (a) the average payoff per agent and (b) the ratio of each strategy over generation when each agent repeatedly plays the public good game with voluntary participation in Table 2. The payoff parameters in Table 2 are set as as follows: R=9, S=0, T=15, P=3, σ=6. There exit some lucky agents who get the highest payoff and unlucky agents who only receive the lowest payoff at the beginning, but their difference becomes narrow and after the 12 generation, the average payoff is increased to 8.9. Three strategies coexist at the beginning but almost all agents gradually become to choose $S_1$ (cooperate), which is a Pareto-efficient strategy.

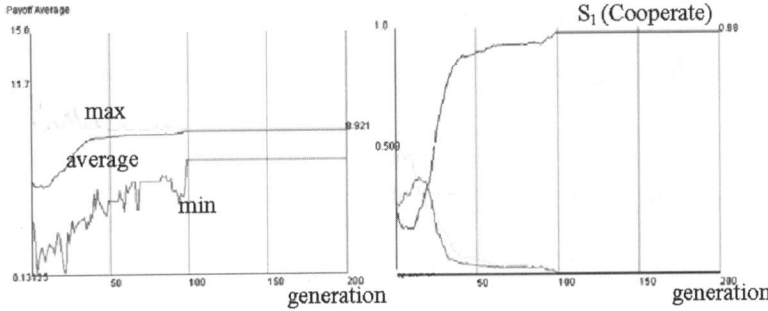

**Fig. 4.** Simulation results with the payoff matrix with loners (R=9, S=0, T=15, P=3)

## 6.2 Dilemma Game with Meta-Strategy

Figure 5 shows (a) the average payoff per agent and (b) the ratio of each strategy over generation when each agent repeatedly plays the Dilemma Game with the meta-strategy in Table 1. The payoff parameters in Table 1 are set as R=3, S=0, T=5, P=1. There exit some lucky agents who get the highest payoff and unlucky agents who only receive the lowest payoff at the beginning, but their difference becomes narrow and after the 50 generation, the average payoff is increased to 2.98. Three meta-strategies coexist at the beginning but almost all agents gradually become to choose $S_3$ (TFT), which is a Pareto-efficient strategy.

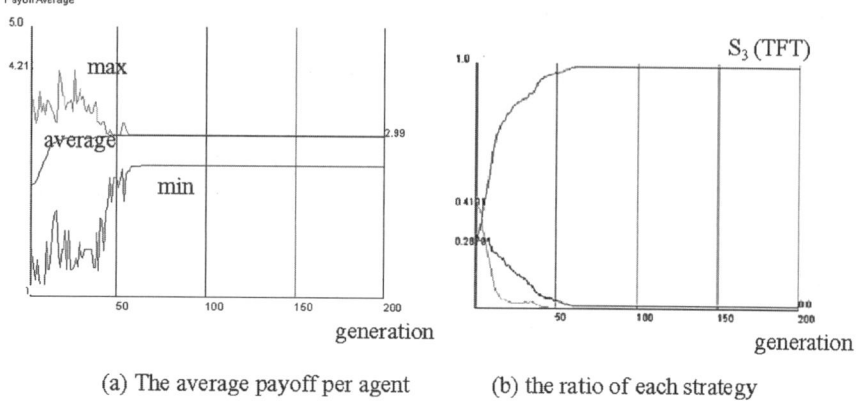

(a) The average payoff per agent          (b) the ratio of each strategy

**Fig. 5.** The state diagram of play between two agents with the same rule  Agents who have the behavioral rules (a) Agents who have the same behavioral rules, (b) Agents who have the different behavioral rules

## 6.3 Implications of Simulation Results

Hauert et al show the importance of volunteering in public good games [9]. Such optional participation is implemented by considering a third strategic type, the loners. Voluntary participation can foil exploiters and overcome the social dilemma. This simple but effective mechanism does not require sophisticated strategic behavior or cognitive abilities but operates efficiently under full anonymity. Thus, it provides an escape hatch out of some social traps. Populations where cooperators abound are prone to exploitation and invasion by defectors but once defectors increase, the loners' option becomes increasingly attractive and as soon as loners dominate cooperation: cooperator beats loner, loner beats defector, and defector beats cooperator. Therefore rock-paper-scissors-type cyclic dominance of strategies can arise whenever the participation in prisoner's dilemma type interactions is voluntary.

Imhof et al also investigate the dilemma game with the three meta-strategies and derive the following conclusions. (i) ALL-C is dominated by ALL-D, (ii) TFT is dominated by ALL-C. (iii) If the average number of rounds exceeds a minimum value, then TFT and ALL-D are bi-stable. Despite ALL-D being the only strict Nash equilibrium in the original dilemma game, they observe that evolutionary oscillations among all three meta-strategies. Therefore the population cycles from ALL-D to TFT, and TFT to ALL-C and back to ALL-D, and it becomes a Rock-Paper-Scissors-type cyclic game.

Strategic environments are often characterized by a tension between individual and collective rationality. A recurring theme in the literature is to understand the conditions under which the potential gains of cooperation manage to override the drawing power of converging to an inefficient situation. Simulation results concerning conditions that facilitate the emergence of Pareto-efficient outcomes have important practical implications. The payoff matrices in Table 2 and Table 3 are examples of games with a Pareto-dominated Nash equilibrium. The underlining game with the payoff matrix in Table 2 has $(S_1, S_1)$ as a Pareto efficient outcome, and the payoff matrix in Table 3 has multiple Pareto efficient outcomes: $(S_1, S_1)$, $(S_1, S_3)$ $(S_3, S_1)$, and $(S_3, S_3)$. Then why is the Pareto efficient outcome $(S_3, S_3)$ is selected among them. If the other agent is choose three strategies with the equal probability in Table 3, the expected payoff of $S_3$ is the highest compared with the other two strategies, therefore the Pareto efficient outcome $(S_3, S_3)$ also risk dominates the other Pareto efficient outcomes.

From the simulation results in the previous two sections, we can conclude that socially intelligent behavior of realizing a collectively desirable outcome is emerged over the networked agents.

# 7 Social Rules Emerged over Networked Agents

Self-interested agents are often faced with the dilemma of acting in their own interest or pursuing a more cooperative course of action. Strategic environments are often characterized by a tension between individual and collective rationality. Social dilemmas are well-known formalizations of such situations. A recurring theme in the literature is to understand the conditions under which the potential gains of cooperation manage to override the drawing power of the Nash equilibrium. A challenging task for researchers is to identify conditions under which agents are more cooperative than the Nash equilibrium based on the assumption of self-interested agents would predict.

Many laboratory experiments and field observations indicate that humans are social animals who take a strong interest in the effects of their actions on others and whose behavior is not always explained by simple models of selfish behavior. Reciprocity and the presence of other behavioral rules can support a great deal of social intelligent behavior [13]. The simulation results in the previous section implicate that we have a tools for examination how socially intelligent behavior evolves in a society of self-interested agents that begins in an amorphous state where there is no established common behavioral rules and individuals only rely on hearsay to determine what to do.

Many spheres of social interactions are governed by social norms such as reciprocity and equity. Social norms are self-enforcing patterns of social behavior. It is in everyone's interests to conform given the expectation that others are going to conform. It is a rule of the action choice that assigns a rule to each agent that is an optimal in the sense no one has an incentive to deviate from it. Although social norms can potentially serve useful constructs to understand human behavior, there is little theory on collective construction of social norm.

Epstein and Axtell extended the literature on the evolution of norms with an agent-based model [3]. In their model, agents learn how to behave (what norm to adopt), but they also learn how much to think about how to behave. The point of their model is that many social norms or conventions have two features of interest. First, they are self-enforcing behavioral regularities. But second, once entrenched, we conform without thinking about it. Indeed, this is one reason why social norms are useful; they obviate the need for a lot of individual computing.

In this paper, I study the collective construction process of social norms as the wisdom of networked agents by focusing on the relation between micro and macro levels of constraints on the evolution of socially intelligent behavior. Collective evolution assumes that successful behavioral rules spread by imitation or learning by the agents. The approach of collective evolution is very much at the forefront of the general topics of designing desired collectives in terms of efficiency, equity, and sustainability. Further work will need to examine how collective evolution across the complex socio-economical networks leads to emergent effects at higher levels. After collective construction of social norms, there is no need to assume a rational calculation to identify the effective behavioral rule.

# 8 Conclusion

Strategic environments are often characterized by a tension between individual and collective rationality. Social dilemmas are well-known formalizations of such and they are typical examples of games with a Pareto-dominated Nash equilibrium. A recurring theme in the literature is to understand the conditions under which the potential gains of cooperation manage to override the drawing power of the Nash equilibrium.

In the paper we present results from computer simulations to show that networked evolving agents are likely to foster social interactions where individual self-interest is consistent with behavior that maximizes the social welfare. Social interaction in such network structure is best modeled as a repeated game. In repeated games, where an agent's actions can be observed and remembered by other agents, almost any pattern of individual behavior, including behavior that maximizes the collective payoff, can be sustained by social norms that include obligations to punish norm violations by others. Where many equilibria are possible, collective construction of social norms is likely to play a major role in determining Pareto-efficient equilibrium will obtain.

We analyzed the emergence of socially intelligent behavior when evolving agents are networked in social spaces. Our problem is to explain how such socially intelligent behaviour could have evolved, given that natural selection operates at the individual level. The framework of collective evolution distinguishes from the concept of co-evolution in three aspects. First, there is the coupling rule: a deterministic process that links past outcomes with future behavior. The second aspect, which is distinguished from individual learning, is that agents may wish to optimize the outcome of the joint actions. The third aspect is to describe how a coupling rule should be improved with the criterion of performance to evaluate how the rule is doing well.

The performance assessment at the individual levels gradually evolves, in order for the agent to act in accordance with the behaviors of her neighbors. Social norms are not merely the union of the local behavioral rules of all agents, but rather evolve interactively, as do the local behavioral rules of the agents. In an evolutionary approach, there is no need to assume a rational calculation to identify the best behavioral rule. Instead, the analysis of what is chosen at any specific time is based upon an implementation of the idea that effective behavioral rules are more likely to be retained than ineffective ones.

# References

1.  Adamatzky, A. (2005). Dynamics of Crowd-Minds, World Scientific.
2.  Axelrod, R., "The Complexity of Cooperation", Princeton Univ. Press, 1997.
3.  Axtell, R and Epstein,M, and Young, P. (2001). The emergence of classes in a multi aget bargaing model in. Social Dynamics, Durlauf, N. and Young, P (eds). Brookings Institution Press, pp. 191-211.
4.  Bergstrom, T. (2002) Evolution of Social Behavior: Individual and Group Selection, Journal of Economic Perspectives, Volume 16, pp. 67-88.
5.  Boyd, R., & Richerson, P. J. (1985) *Culture and the evolutionary process*. The University of Chicago Press.
6.  Dick, S, Krambeck, M, and Milinski, S (2003) Volunteering leads to rock-paper scissors dynamics in a public goods game, Nature. Vol. 425, pp. 390-393.
7.  Ellison, G. (1993). Learning local interaction, and coordination, Econometrica, 61, pp. 1047-1071.
8.  Epstein, J. M. and Axtell, R. (1996). Learning to be thoughtless: Social norms and individual computation, Working paper no. 6, Brookings Institution.
9.  Hauert, C., De Monte, S., Hofbauer, J. and Sigmund, K. (2002) *Volunteering as Red Queen Mechanism for Cooperation in Public Goods Games*, Science, Vol. 296 1129-1132.
10. Hauert, C.and Szabo. (2005) Game Theory and Physics, *Am. J. Phys.73*, pp. 405-411.
11. Imhof, L. Fudenberg, D and Nowak, M. (2005) Evolutionary Cycles of Cooperation and Defection, *PNAS,* Vol.102, pp. 10795-10800.
12. Kuperman and Abramson (2001), Social games in a social network, Phys. Rev. E 63.
13. McMahon,C. (2003). Collective Rationality and Collective Reasoning, Cambridge University Press.
14. Namatame, A., Sato, N., Murakami Y.(2004). *Co-evolutionary Learning in Strategic Environments*, World Scientific, pp. 1-19.
15. Namatame, A. (2006) Adaptation and Evolution in Collective Systems, World Scientific.
16. Nowak, A (2006). Emergence of Social Complexity form Simple Interactions among People, memo.
17. Nowak, M. A. and Sigmund, K. (2004). Evolutionary dynamics of biological games, Science, 303, pp. 793-799.
18. Semmann, H, Krambeck, H and Milinski, S: (2003) *Volunteering leads to rock-paper-scissors dynamics in a public goods game*, Nature vol. 425, pp. 390-393, 2003.

19. Skyrms, B and Pemantleá, R. (2000) A dynamic model of social network formation, PNAS, Vol. 97, pp. 9340-9346.
20. Surowiecki, K, *The Wisdom Of Crowds*, Anchor Books, 2004.
21. Roughgarden, T. (2005). Selfish Routing and the Price of Anarchy, The MIT Press.
22. Tumer, K & Wolpert, D.(Eds) (2004). *Collectives and the Design of Complex Systems*, Springer.
23. Verplanken B, Aarts H (1999). Habit, attitude, and planned behavior. European Review of Social Psychology, Vol.10, pp. 101-134.
24. Watts, D. and Strogatz, H. (1998). Collective dynamics of small-world networks, Nature, 393, pp. 440-442.
25. Young, H. P. (2005). Strategic Learning and Its Limits, Oxford Univ. Press.

# Using an agent based simulation to evaluate scenarios in customers' buying behaviour

Filippo Neri

DSTA, University of Piemonte Orientale,
Via Bellini 25/g, 15100 Alessandria (AL), Italy `filipponeri@yahoo.com`

An agent based tool for analyzing markets behavior under several rates of information diffusion is described. This methodology allows for the study of tradeoffs among several variables of information like product advertisement efforts, consumers' memory span, and passing word among friends in determining market shares. Insights gained by using this approach on an hypothetical economy are reported.

## 1 Introduction

The diffusion of an Internet based economy, that includes even the less valuable transactions, is day by day more evident. The existing information infrastructure has allowed the exploitation of new methods to contract the purchases of goods and services, the most notable of which is probably the agent mediated electronic commerce [Kephart et al., 1998, Maes, 1994]. In this economy, autonomous agents become the building block for developing electronic market places or for comparing offers across several seller's websites (shopbots) [Maes, 1994, Rodriguez-Aguilar et al., 1998, Lomuscio et al., 2001]. The possibilities offered by the new shopping environment results in the consumer adopting a (possibly) completely new decision making process to select which product to buy among the available offers.

Our aim is to use an agent-based market place to qualitatively simulate the diffusion of products' awareness across the Internet and its impact on customer choices. Another important motivation in our decision to adopt an agent based simulation framework is that we aim to study how individual history and limitations impact on group dynamics. As many commercial scenarios could be selected, we chose to model a simple commercial interaction. Different groups of consumers have to choose one product between a set of perfect substitutes that differ in price, advertised lifestyle associated with the product and the advertising effort to initially penetrate the market. Our objective is the to understand how a sequence of repeated purchases is affected by the trade

F. Neri: *Using an agent based simulation to evaluate scenarios in customers' buying behaviour*,
Studies in Computational Intelligence (SCI) **56**, 177–188 (2007)
www.springerlink.com © Springer-Verlag Berlin Heidelberg 2007

off among the previous variables, the consumers' desires and limits, and the diffusion of the awareness about the existing products. The modelling ultimate goal would be to capture the common experience of choosing, for instance, among alternative brands of Italian Pasta packages displayed in the webpage or on the physical shelf of our grocery store.

Our research scope is in between the investigation of consumer decision making [Bettman, 1979, Guadagni and Little, 1983] and the study of electronic based economies of software agents, shopbots economies for short, [Maes, 1994, J. G. Lynch and Ariely, 2000]. In the following we describe the relationships of our work with both fields. Consumer decision making has received attention from a number of different research fields including psychology and the quantitative modelling communities. Psychological research aims to understand the reasons underlying the decision making process, whereas quantitative modelling community uses a variety of techniques from statistics, machine learning, and software agents to quantitatively model the factors that are involved in the decision making process. A summary of the most relevant works in the area is reported.

Cognitive investigation of consumer decision making. Bettman [Bettman, 1979] and Bettman et al. [Bettman et al., 1988] investigate how consumers decide what product to buy. They propose that consumers have a limited (information) processing capability, they act in order to satisfy a need, and usually do not have a well defined set of preferences to be used in product selection. Instead they construct them using a variety of strategies which depend on the situation at hand. Bettmanns work aims to make explicit the cognitive framework (i.e. state its underlying constraints) used by the consumer to then build the mental model used when deciding what to buy. Practically, a ready-to-use model of the consumers able to provide quantitative indication cannot be immediately derived by Bettmans work. From our perspective, Bettman et al.'s work show that consumers engage in a mental process consistent with weighted adding in less emotional buying tasks (i.e. buying groceries vs. buying an house). Also they show that choice processes can be selective (some products are filtered out), comparative (among the filtered remaining products) and influence the items stored in the consumer memory. Hoyer [Hoyer, 1988] proposes that consumers used different decision making strategies, not only because of individual differences, but depending on the high/low involvement (i.e. risk and/or emotional impact of purchasing the product) they feel toward the product category. Out-of-the-store decision making strategies, meaning that the consumer has already decided which product brand she will buy before reaching the store, are considered and empirical test of these hypothesis are carried out. [Wright, 1975] proposes a framework to analyse the strategies used by consumers to choose between alternative products. Strategies are evaluated by volunteers, participating in a psychological experiment, in term of the easiness to be remembered and applied at the moment of purchase. No attempt to model real consumers in a real store is however address.

Quantitative analysis of consumer decision making. Guadagni and Little [Guadagni and Little, 1983] uses a multinomial logit model of brand choice to predict the share of purchases by coffee brand and (package) size. The model has been developed by using data collected through optical scanning of products at check out in supermarkets. This work is similar to our approach but we differentiated in term of technology used (software agents vs multinomial logit model), of the explicit modelling of consumers types and preferences and because we want to take into account information exchange among consumers between repeated purchases. Currim et al. [Currim, 1988] show how to use decision trees to represent the process consumers use to integrate product attributes when making choices. Models of the individual consumer or of consumers segments are studied in the case of choosing among alternative brands of coffee. The work is exploratory in nature and no attempt to calculate product market shares is made. Smith and Brynjolfsson [Smith et al., 2001] shows how data collected by an internet shopbot, a web-engine that compare offers for the same product but from different retailers, can be used to analysed consumer behaviours and their sensitivity to issues like products price, shipping conditions and brand name of the retailers. Degeratu et al.[Degeratu et al., 2000], for instance, explores how correlations between brand name, price and sensory attributes influence consumer choice when buying on-line or off-line. They conclude that prices for products in the same category are not the main drivers for consumer decision both when shopping on-line and off-line. Instead a combination of price together with additional information, intrinsically product dependent, is used buy the customer to take a decision. Degeratu et al.'s approach is based on a stochastic model of the consumers whose parameters are inferred by real data. In this case the consumer model is a probabilistic equation where random variables account for the variation observed in the data. Our approach, instead, aims to characterise typologies of consumer through an explicit definition of the key drivers under the buying decision.

Lynch and Ariely [J. G. Lynch and Ariely, 2000] try to understand the factors behind purchases made in a real world experiment of wine selling across different retailers' websites, but no consumer model is produced.

Software agents modelling of market environments. Rodrguez-Aguilar et al. [Rodriguez-Aguilar et al., 1998] explore how to use agents to define and study (electronic) auction markets and proposed that such competitive situation constitute a challenge for software agents research in the area of agent architectures and agent based trading/negotiation principles. We agree with them and we plan to evaluate some of their idea to enrich the communication part of our simulation approach. Sophisticated interactions between agents, negotiation strategies, in market-like environments are also being studied, see for instance [Rocha and Oliveira, 1999, Viamonte and Ramos, 1999, Sierra et al., 1998]. Our approach relies on a simple exchange of information for the moment. Finally we are not aware of any other agent based approach, other than ours, trying to capture the complexity of the consumer decision making process at the buying location. In term of relationships of our work with research in the area of shopbots economies, we not that some researchers take a very long term view about the ecommerce phenomena envisioning economies of shopbots [Kephart et al., 1998, Maes, 1994, Rodriguez-Aguilar et al., 1998]. For instance, Kephart et al. [Kephart et al., 1998] try to model large open economies of shopbots by analysing an economy based on information filtering and diffusion towards targeted shopbots (customers). Quite differently, we try to capture the commercial phenomena in a more near future where customers are human beings with their intrinsic limit in information processing, having the need to trust the bought product and to feel supported, and reassured about their purchasing choice as their best possible choice. We share with Kephart et al. the desire to analyse and understand how the information flow can affect such economy. Indeed our aim is to use an agent-based market place to qualitatively simulate the diffusion of products' awareness across the Internet and its impact on customer choices [Neri, 2001].

The Minority Game [Brian, 1994] is an agent based simulation based on the same philosophy as ours. The Minority Game is a game played by artificial agents with partial information and bounded rationality where the minority wins. For instance, in [Brian, 1994], 100 agents have to independently decide whether to go or not to El Farol's bar knowing that if more than 60 agents are there, the bar will be too crowded to be enjoyable. Each agent thus has to devise a strategy about when to go to the bar depending on his past history of successes or failures. Our approach however differs from the Minority Game in the sense that we stress the importance of experience sharing and communication among agents.

Hales' work [Hales, 1998] also related to ours. Hales explores the relationship between agents, their beliefs about their environment, communication of those beliefs, and the global behaviours that emerge in a simple artificial society. Our work differs however for the domain and the focus toward defining the individual behaviour responsible for an observable macro effect at the level of the whole economy.

To further extend our work, a more sophisticated approach to modeling the electronic market place may have to be selected in order to take into account negotiation protocols or virtual organisation formation as, for instance, described in [Rocha and Oliveira, 1999] or to account for additional brokering agents as describe in [Viamonte and Ramos, 1999]. In the near future, we would like to investigate the emergence of information diffusion strategies by using a distributed genetic algorithm [Neri and Saitta, 1996].

The paper is organised has follow: in section 2 a description of the market place simulation is reported, in section 3 the performed experiments are commented and, finally, some conclusions are drawn.

# 2 The Virtual Market Place

The architecture of the agent based virtual market place is quite simple: one purchaising round after the other, groups of consumers, modelled as software agents, select which product to buy according to their internal status. The internal status takes into account the consumers' preferences for a product and her awareness about the product's benefits and image. This process based description of the buying experience matches what most people experience when selecting among alternative wholemeal breads or milk chocolate bars at the local grocery store [Bettman, 1979]. In the simulator we represent both products and consumers as software agents. A product is a collection of an identifier, a price, an effort to describe its features/benefits on the package, an effort to bound the product to the image of a lifestyle (brand) and an initial advertisement effort to penetrate the market. It is important to note that the scope of this work is to consider products that are substitute one for the others but differ in price or other characteristics. The idea to model products as software agents is new.

A consumer is a (software) agent operating on the market and driven in her purchases by a target price, a need for understanding the product benefits, the lifestyle conveyed by the product brand, and the initial marketing effort put into placing the product in the market. The consumer can remember only a constant number of products (memory limit) for a constant number of rounds (memory duration), and she may share with her friends her opinion about the known products. It is worthwhile to stress that the memory span limits the consumer awareness of the available products. For instance, if a consumer had a memory limit of 3, she would be aware of 3 products at most and she would make her choice only among those three products. A consumer will not remember a product, if its memory has already reached its limit, unless it is better of an already known product thus replacing it. However, round after round, consumers talk to each other and they may review their opinions about the products by updating their set of known products. Our interest lays in forecasting the product market shares (percentage of bought products) on the basis on the previous factors. In order to evaluate the feasibility of our approach, we developed from scratch a basic version of the market place simulator and performed some experimentation under constrained conditions.

In the following the detailed descriptions of both the simulator's architecture and the experimental setting is described. In the simulator, each product is defined by an identifier (Id), a selling price (Price), an effort in describing its benefits on its package, an effort to convey a lifestyle (image), and an effort to initially penetrate the market. As an instance, in the initial series of experiments, all the products prices and characteristics are selected to cover a wide range of significant offers as follow:

Product(Id, Price, Description, Image, InitialAdvertisement)
Product(0, LowValue, LowValue, LowValue, LowValue)
Product(1, LowValue, LowValue, LowValue, HighValue)

Product(2, LowValue, LowValue, HighValue, LowValue)
Product(3, LowValue, LowValue, HighValue, HighValue)
Product(4, LowValue, HighValue, LowValue, LowValue)
Product(5, LowValue, HighValue, LowValue, HighValue)
Product(6, LowValue, HighValue, HighValue, LowValue)
Product(7, LowValue, HighValue, HighValue, HighValue)
Product(8, HighValue, LowValue, LowValue, LowValue)
Product(9, HighValue, LowValue, LowValue, HighValue)
Product(10, HighValue, LowValue, HighValue, LowValue)
Product(11, HighValue, LowValue, HighValue, HighValue)
Product(12, HighValue, HighValue, LowValue, LowValue)
Product(13, HighValue, HighValue, LowValue, HighValue)
Product(14, HighValue, HighValue, HighValue, LowValue)
Product(15, HighValue, HighValue, HighValue, HighValue)

The constants 'LowValue' and 'HighValue' correspond to the values 0.2 and 0.8. The Price, Description and Image parameters are used to evaluate a customer's preference for the product, whereas the InitialAdvertisement parameter defines the initial awareness of the product among the customers. So, for instance, a product defined as Product(x, LowValue, LowValue, LowValue, LowValue) is especially targeted toward price sensitive consumers that do not care about knowing much on the product. And with an initial penetration rate of 0.2, on average, 20% of the consumers are aware of its availability at the beginning of the first buying round. Finally, it is worthwhile to note that, in the above list, odd and pair numbered products differ only because of a different initial advertising effort.

A similar representation choice has been made to represent customers. Four groups of consumers are considered. For the scope of the initial experiments, we concentrate on customers whose target product has a low price but differs in the other features. Consumer groups are represented as follows:

Customer(Price, Description, Image)
Customer(LowValue, LowValue, LowValue) (bargain hunters)
Customer(LowValue, LowValue, HighValue) (image sensitive)
Customer(LowValue, HighValue, LowValue) (description sensitive)
Customer(LowValue, HighValue, HighValue) (image and description sensitive)

Through the selection of target values, we tried to capture the following categories of customers: the bargain hunters, the brand sensitive ones, the package sensitive ones (i.e. are interested in its nutrition values, its composition, its ecological impact, etc.), and those that are both brand and package sensitive. It is important to note that each customer does not necessary known the same products than other consumers because of the individual memory and of the initial random distribution of a product awareness among consumers. During each round, a consumer chooses to buy the product that most closely matches her preferences.

According to Bettman [Bettman, 1979] and [Hoyer, 1988], we approximate the product matching process by means of a weighted average function defined as follows:

$Preference(product) =$
$(max(product.Price, target.Price) -$
$target.Price)^2 +$
$(min(product.Description, Description) -$
$target.Description)^2 +$
$(min(product.Image, target.Image) - target.Image)^2$

The preferred and selected product is the one with the lowest value of the Preference function among the ones known by the customer. Alternative expressions are under study.

Also each customer does not necessarily known the same products than the others because of the different distribution of the products depending on their initial marketing effort. The reported experiments aim to understand the impacts of the following factors in determining the final product market shares: customer preference definition, initial market penetration effort, number of friends in passing the word of known products, and memory limit.

In the initial group of experiments we aimed to investigate some hypothesis on the impacts of the diffusion of product awareness and shift in the consumers' behaviours [Neri, 2001]. The obtained results are promising and confirm the feasibility of the approach. They are however far from being conclusive in term of hypothesis testing. Indeed in order to perform extensive and informative experiments, the virtual market place simulator should be completely re-engineered to facilitate its use and the definition of hypotheses/rules governing the consumers' behaviour.

# 3 Experimental Results

Goal of the experimentation is to show that our tool can capture some of the inherent complexity behind the determination of the product market shares by considering a variety of factors that impact on this economic phenomena. These factors include the customers' expectations for a product, the limited memory span and duration that consumers reserve to remember available products, and the diffusion of the product awareness among consumers by initial advertisement and further passing by word. Value ranges for this variables have been selected accordingly to past experience with consumers behaviour.

All the reported experiments refer to an hypothetical economy and are based on the following basic settings. During each round, 400 consumers (one hundred for each of the four consumer types) select which of the 16 products to buy. Only products that the consumer remembers (i.e. appearing in its memory list) compete for being purchased. The economic process is repeated for 100 rounds. For each experiments, the reported figures are averaged over 3 runs.

As a baseline for evaluating the economic process, we consider the situation where each consumer is fully aware of all the available products since the first round. As all the consumers are oriented towards products with low price but with different characteristics, it is straightforward to calculate that the product market shares stay constant over the 400 rounds and correspond to the values reported in Fig. 1. In the picture, the product's identifiers appear on the x axis, and the market shares on the y axis. Thus for instance, Product 6 will achieve a 9.3% market share. It is worthwhile to note that the product from 9 to 16 have a 0% market share because, in the range from 1 to 8, there exists a product with identical features but with lower price.

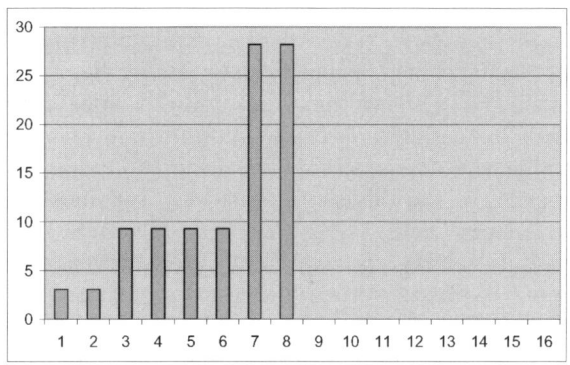

**Fig. 1.** Ideal market share distribution in presence of a perfect product awareness or perfect information flow.

If we were in this ideal situation, every consumer would be able to make the best pick among the available products. Unfortunately, in the real world, full knowledge about the available choices is not common and product awareness is the results of a variety of factors including advertisement, passing by word among friends and memory capacity. The impact of these factors on the product market shares is taken into account in the following experiments.

Let us consider the case where consumers do not have any friends or do not talk about products to friends (average number of friends or avgf = 0), they can remember only 2 products at the time (memory limit or ml = 2), and they remember each product for 20 rounds unless either they keep buying it or they are told about by their friends. The initial (end of round 1) and final market shares (end of round 100) appear in Fig. 2.

It appears that the initial and final market shares are very alike and that the higher the effort in penetrating the market the better the market share (compare odd and even numbered products). The market share distribution is biased toward low priced product, this is to be expected given the customers' preferences. But, still, some high price products achieve a significant portion

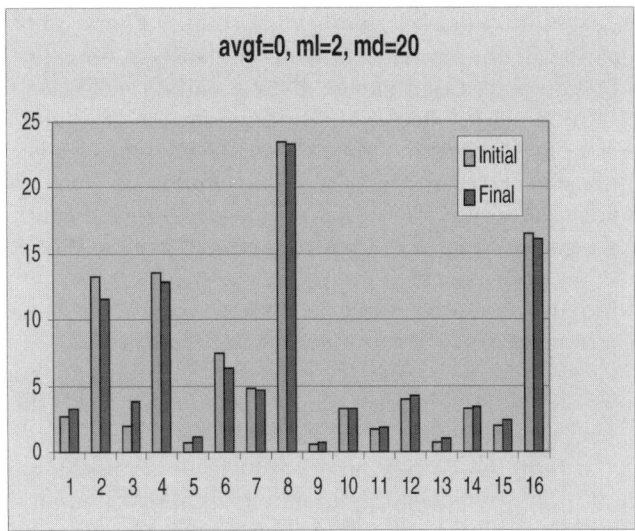

**Fig. 2.** Product market shares in the case of consumers not talking to their friends about their shopping (avgf=0), remembering at most 2 products (ml=2) and with memory duration of 20.

of market because of the limited memory span of the consumers that would prevent him to compare and choose among more alternatives.

If we alter the previous scenario just by increasing the number of friends to 20, we obtain quite a different distribution of market shares, Fig. 3. The

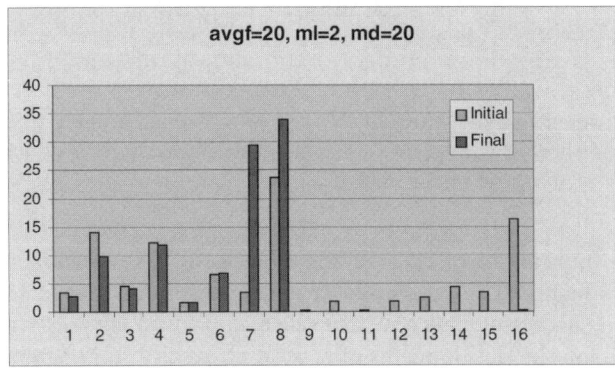

**Fig. 3.** Product market shares in the case of consumers talking to about 20 friends about their shopping (avgf=20), remembering at most 2 products (ml=2) and with memory duration of 20.

pattern of the initial market shares is, of course, similar to that of the previous scenario but the final shares tends to converge towards the ideal ones. This can be interpreted that having many friends or collecting many opinions among the same market does actually empower the customer in making the best selection. It is interesting to note that the only initial advertisement cannot compensate for the further product comparisons communicated among the consumers. However, the initial product advertising effort results in the consumers remembering and, then, choosing the more advertised products among the low priced ones.

An alternative scenario would be to keep an average number of friends equal to 0, but increase the consumer memory limit to 12.

In this case, the initial and final distribution would look alike and tend to converge to the ideal market shares distribution but a bias toward the products investing in the initial advertising is evident.

Finally, if both the average number of friends (avgf=20) and the memory limit (ml=12) increase, then the initial and final distribution differ, the final one most closely matching the ideal ones, Fig. 4. Comparing the initial and

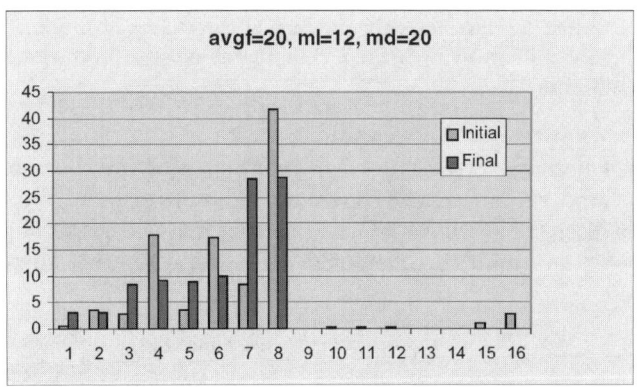

**Fig. 4.** Product market shares in the case of consumers talking to about 20 friends about their shopping (avgf=20), remembering at most 12 products (ml=12) and with memory duration of 20.

final distributions of market shares it appears that exchanging information about products with friends and remembering a number of them is the key to make a successful choice in this scenario. Indeed this observation is at the very base for the development of several strategies to deal with comparative on-line shopping.

## 4 Conclusion

An agent based methodology and tool to study market behaviors under several conditions of information diffusion has been described. The reported experimentation, in the context of an hypothetical economy, shows how this approach can be used to analyse and visualize market shares resulting after many complex information-based interactions among economic agents.

Concerning electronic shopping and, especially, comparative shopping engines, the reported experiments show the significance of exchanging information among economic agents. Indeed, this is the key to make good/bad buying choice. Obviously, buyers and sellers regards each choice from a different perspective. This observation and this approach can help the development of novel marketing strategies in the comparative on-line shopping environment.

# References

[Bettman, 1979] Bettman, J. (1979). *An information processing theory of consumer choice*. Addison Wesley, Reading, MA (USA).

[Bettman et al., 1988] Bettman, J., Luce, M., and Payne, J. (1988). Constructive consumer choice processes. *Journal of Consumer Research*, pages 187–217.

[Brian, 1994] Brian, A. W. (1994). Inductive reasoning and bounded rationality. *American Economic Review*, pages 406–411.

[Currim, 1988] Currim, I. (1988). Disaggregate tree-structured modeling of consumer choice data. *Journal of Marketing Research*, pages 253–265.

[Degeratu et al., 2000] Degeratu, A. M., Arvind, R., and Wu, J. (2000). Consumer choice behavior in online and traditional supermarkets: The effects of brand name, price, and other search attributes. *International Journal of Research in Marketing*, pages 55–78.

[Guadagni and Little, 1983] Guadagni, P. and Little, J. (1983). A logit model of brand choice calibrated on scanner data. *Marketing science*, pages 203–238.

[Hales, 1998] Hales, D. (1998). An open mind is not an empty mind - experiments in the meta-noosphere. *The Journal of Artificial Societies and Social Simulation (JASSS)*.

[Hoyer, 1988] Hoyer, W. (1988). An examination of consumer decision making for a common repeat purchase product. *Journal of Consumer Research*, pages 822–829.

[J. G. Lynch and Ariely, 2000] J. G. Lynch, J. and Ariely, D. (2000). Wine online: search costs and competiotion on price, quality and distribution. *Marketing Science*, pages 1–39.

[Kephart et al., 1998] Kephart, J. O., Hanson, J. E., Levine, D. W., Grosof, B. N., Sairamesh, J., Segal, R., and White, S. R. (1998). Dynamics of an information-filtering economy. In *Cooperative Information Agents*, pages 160–171.

[Lomuscio et al., 2001] Lomuscio, A., Wooldridge, M., and Jennings, N. R. (2001). A classification scheme for negotiation in electronic commerce. In *AgentLink*, pages 19–33.

[Maes, 1994] Maes, P. (1994). Agents that reduce work and information overload. *Communications of the ACM*, pages 31–40.

[Neri, 2001] Neri, F. (2001). An agent based approach to virtual market place simulation. In *Congresso dell'Associazione Italiana Intelligenza Artificiale 2001 (AIIA01)*, pages 43–51.

[Neri and Saitta, 1996] Neri, F. and Saitta, L. (1996). Exploring the power of genetic search in learning symbolic classifiers. *IEEE Trans. on Pattern Analysis and Machine Intelligence*, PAMI-18:1135–1142.

[Rocha and Oliveira, 1999] Rocha, A. P. and Oliveira, E. (1999). Agents advanced features for negotiation in electronic commerce and virtual organisations formation process. *Agent Mediated Electronic Commerce - An European Perspective*, LNAI 1991.

[Rodriguez-Aguilar et al., 1998] Rodriguez-Aguilar, J. A., Martin, F. J., Noriega, P., Garcia, P., and Sierra, C. (1998). Towards a test-bed for trading agents in electronic auction markets. *AI Communications*, 11(1):5–19.

[Sierra et al., 1998] Sierra, C., Jennings, N., Noriega, P., and Parsons, S. (1998). A framework for argumentation based negotiation. In *Intelligent Agents IV*, volume LNAI 1365, pages 177–192, Vienna, Austria. Springer-Verlag.

[Smith et al., 2001] Smith, M. D., Bailey, J., and Brynjolfsson, E. (2001). Understanding digital markets: review and assessment. *Draft available at http:ecommerce. mit.edupapersude*, pages 1–34.

[Viamonte and Ramos, 1999] Viamonte, M. J. and Ramos, C. (1999). A model for an electronic market place. *Agent Mediated Electronic Commerce - An European Perspective*, LNAI 1991:3–28.

[Wright, 1975] Wright, P. (1975). Consumer choice strategies: symplifying vs optimizing. *Journal of Marketing Research*, pages 60–67.

# How to Form Stable and Robust Network Structure through Agent Learning — from the viewpoint of a resource sharing problem

Itsuki Noda[1] and Masayuki Ohta[1]

Information Technology Research Institute
National Institute of Advanced Industrial Science and Technology
{i.noda,m-ohta}@aist.go.jp

**Summary.** In this article, we consider forming a social structure of agents as a resource sharing problem among the agents, and propose a method to adapt learning parameters for multi-agent reinforcement learning (MARL) to stabilize the social structure. In general, learning parameters like temperature of Boltzmann softmax functions or $\epsilon$ in $\epsilon$-greedy methods are important control factors for balancing exploration and exploitation in reinforcement learning. In addition, controlling global information for MARL is also a key issue to balance speed and convergence of the learning. We apply a combination of algorithm of an exploration factor of the agent learning and a moderation factor of the moderated global information based on WoLF principle. In the method, ratio of exploration is decreased when the best agent policy gets more than expected reward, while the ratio is increased otherwise. By this method, the agent can get a suitable policy robustly and also the social structure is stabilized quickly. We conduct several experiments to show the stability and robustness of the proposed method.

## 1 Introduction

When we consider a social structure as an network of information flow, there are two possible aspects: 'robust connection' and 'management control'. From the view point of 'robust connection', a fully connected network is the best organization, because agents can choose any connections dynamically according to changes of environments. It is, however, ineffective from the view point of 'management control, because there are many duplications of management flows in the network especially in the case where a certain resource like manpower is bounded. Generally, a stable tree structure is the best organization from the view point of the management under static environments.

When we focus on the process to form social structure, on the other hand, it is not trivial for selfish learning agents to form a network of the stable tree

I. Noda and M. Ohta: *How to Form Stable and Robust Network Structure through Agent Learning— from the viewpoint of a resource sharing problem,* Studies in Computational Intelligence (SCI) **56**, 189–201 (2007)
www.springerlink.com

structure. Especially, simultaneous learning of multiple agents causes unstable environmental change for each other. This issue becomes clear when we consider forming the social network as a resource sharing problem in which each agent seeks the best connection with another agent who will provide the best benefit. Because the ability of each agent is limited, the connections should be distributed suitably among the agent society.

Resource sharing problems (RSP)in multi-agent systems are a fundamental subject both in information technologies and in social science. In previous works, this subject has been formalized as a optimization problem, in which we focused on mechanism to control agents' behaviors to reach a social optimum where total effectiveness/benefits of agents are maximized [9, 4]. On the other hand, RSPs can be formalized as a stability issue, in which we focus on a way to manage agents and their environments to stabilize total behaviors of agents. For example, when all agents are selfish and have the same knowledge about expected benefits for each connection, they tend to choose an identical connection. As a result, there occur serious congestion. If this undesirable situation are reported to all agent and the agent update their knowledge in the same way, they tend to choose another identical connection in the next turn so that another congestion occurs again. Such changing congestion occurs cyclically or irregularly, and brings social ineffectiveness. We call such kind of phenomena as *oscillation problem*. If we can stabilize agents' choice, we will be able to make a plan to reinforce the resources to provide more benefit to agents. We can see similar oscillation problems in several domains like choice of attractions in theme parks, load-balancing in computer-networks, and so on. In order to avoid such situation, several mechanisms are proposed [11, 10, 6].

The important point of the problem as a stability issue is how to control 'selfish-ness' of agents: When we suppose agents are selfish, there are no way for each agent to select a bad choice. While existing works for this problem use heterogeneity and priorities of agents, randomness, or market mechanisms, we are focusing Nash equilibrium, in which all agents are satisfied because each of them believes its choice is the best. Satisfying all agents is important in open systems like social systems because we can not suppose that all agents are well-mannered.

In order to attack the stability issue, we are proposing a new approach in which additional global information about benefits of resources is provided to selfish and learning agents [7]. By using the global information, which consists of a collection of moderated benefits of all resources, each agent can reach Nash equilibrium and stabilize its behavior.

In this paper, we consider the process to form a social network as resource sharing problem, and propose a meta-level mechanism to control the additional information for more general and dynamic environment. In the rest of this paper, we formalize the resource sharing problem in section 2 and propose a framework to control agent learning based on WoLF principle in section 3. The propose method is evaluated through several experiments in section 4.

## 2 Resource Sharing and Learning Agents

### 2.1 Resource Sharing Problem

An RSP is defined as follows: Suppose that there exists a set of resources $R = \{r_1, r_2, \cdots, r_n\}$ and a set of agents $A = \{a_1, a_2, \cdots, a_m\}$. In every discrete time step, the following procedure is executed:

S.1 Each agent $a_i$ in $A$ chooses a resource $r_j$ in $R$ according only to the agent's policy.

S.2 Each agent $a_i$ who chooses resource $r_j$ receives a utility which is calculated by the utility function $(U_j)$ of resource $r_j$ and the number of agents $(n_j)$ who choose the resource $r_j$.

Here, "according only to the agent's policy" means all agents are selfish and introverted: A decision of each agent is never influenced by other's simultaneous behaviors or thinking. We also suppose the policy is without memory, that is, each decision is not influenced by a history of past decisions of the agent. Therefore, the policy can be defined as a probabilistic function $\pi_i(r_j)$, which indicate a probability that agent $(a_i)$ chooses resource $r_j$.

Because the problem is resource sharing, the utility of a resource for an agent should decrease when more agents choose the resource. Therefore, we suppose that the utility function $U_j$ of resource $r_j$ is a monotonically decreasing function, that is,

$$\frac{dU_j(n_j)}{dn_j} < 0 \quad (n_j > 0) \tag{1}$$

In addition, we suppose that each agent adjusts its policy using reinforcement learning based on its experiment:

S.3 Each agent $a_i$ changes its policy $\pi_i(r_j)$ according to the utility $U_j(n_j)$ the agent gets.

Because of the simplicity, we use the following framework as the learning mechanism: Each agent $a_i$ has its own estimated utility of resource $r_j$, which is denoted as $V_i(r_j)$. When the agent receives a utility from $u_j$ from resource $r_j$, the agent modifies its estimation as follows:

$$V_i(r_j) = (1 - \alpha)V_i(r_j) + \alpha u_j, \tag{2}$$

where $\alpha$ $(0 < \alpha < 1)$ is a learning rate. The policy of agent $\pi_i(r_j)$ is calculated from $V_i(r_j)$ using Boltzmann softmax function as follows:

$$\pi_i(r_j) = \frac{e^{V_i(r_j)/T}}{\sum_{r_k} e^{V_i(r_j)/T}}, \tag{3}$$

where $T$ $(T > 0)$ is a temperature parameter

## 2.2 Moderated Global Information

One of the problems of multi-agent reinforcement learning is the number of learning example and the variation of situations. Generally, each agent can be considered as an environment for other agents. For example in RSP, when an agent change its decision from resource $r_j$ to resource $r_k$, other agents who are choosing resource $r_j$ and $r_k$ get more and less utilities, respectively, even they do not change their policy. This means that the number of situations increases drastically when the number of agents increase. On the other hand, it is difficult to increase the number of learning example especially in actual problems.

In order to overcome this problem, we have been proposing a method to utilize moderated global information (MGI) [7]. In MGI method, in addition to learning by self-experiments eq. 2, each agent adjusts estimated utilities of resources that the agent does not choose as follows:

$$V_i(r_k) = (1 - \alpha')V_i(r_k) + \alpha' U_k(n_k + l), \tag{4}$$

where $\alpha'$ $(0 < \alpha' < 1)$ is a learning rate, and $l$ $(l \geq 1)$ is a moderation factor.

If $l = 0$, eq. 4 means that each agent can learn from other's experiences immediately without modification. This kind of methods are used in various situations of multi-agent learning like imitation learning[8]. However, the direct usage of other's experiences causes a serious problem in an RSP, that is, all agents learn an identical estimated utility by sharing the same experiments with each other, so that they become to choose an identical resource simultaneously. As a result, an oscillation problem occurs.

On the other hand, if we set $l = 1$, the behaviors of learning agents changed drastically. In [7], we theoretically showed that policies of learning agents fall into a Nash equilibrium and is stabilized at the equilibrium. Using this effect, we demonstrated that the moderated information can reduce the adverse effect caused by rumors in selfish agents shown in the previous paragraph.

## 2.3 Exploration and Exploitation

The MGI method still has an open issue that it is not guarantee the convergence of agents' policies into Nash equilibrium especially in the case of dynamic and swiftly-learning agents. This issue is a kind of a balancing problem between exploration and exploitation in multi-agent learning [3]. One solution for this problem is to change several learning parameters through time. For example, [5] applied simulated annealing technique into reinforcement learning to balance between exploration and exploitation. In their method, a temperature parameter like one used in eq. 3 drops toward zero through time. Finally, each agent gets a deterministic policy in which there are no randomness in the choice of the agent. While such deterministic behavior is desirable from the view point of the stability, the monotonic dropping method may cause a serious oscillation problem when an environment changes after the conversion. Figure 1 shows a typical oscillation problem when we use a monotonic dropping method in learning. This figure shows the changes of the number of agents who choose one of three resources. In the beginning(left in the graph), agents choose resource relatively randomly but learn to acquire a Nash equilibrium. However, the agents start to wander among resource at step 100 when the environment (utility functions of resources) is changed. After that the wandering never stop but continue oscillation.

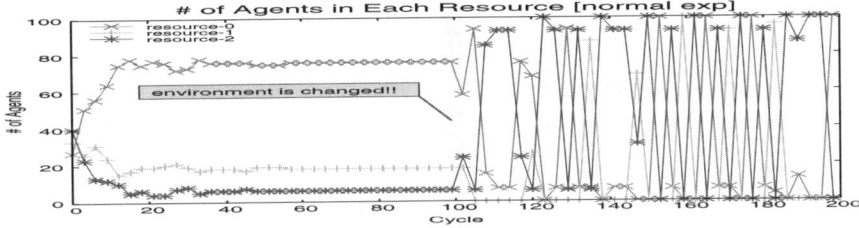

**Fig. 1.** Typical Oscillation Problem in Resource Sharing

Another interesting idea on this convergence issue is WoLF (Win or Learn Fast) principle [1, 2]. In this principle, learning agents use different learning parameters for winning and losing phases: If a agent is losing, a set of parameters is used to enforce the learning faster than in the winning phase. Using this principle, Bowling et al. show that learning agents can avoid oscillation problem and reach equilibrium.

In this paper, we generalize the concept of WoLF principle and derive a framework to change learning parameters by which we control the learning process to converge into an equilibrium.

## 3 Proposed Method

### 3.1 Abstracted WoLF Principle

It is difficult to apply the WoLF principle directly to RSP because of two reasons. First, WoLF supposes that Nash equilibriums exist not in pure strategies but in mixed strategies, while RSP should have a equilibrium in pure strategies from the view point of the stability of agents' behaviors. Second, the original WoLF method supposes that each agent learns Q values (estimated utilities) and probabilities of actions separately, while we suppose the agent use Boltzmann softmax function to determine the probabilities.

To overcome this difficulty, we try to generalize and abstract the concept of WoLF principle from the view point of 'exploration and exploitation', which is a main issue of RSP as shown in the previous section. In WoLF, an agent

(1) Let $\alpha \in (0, 1]$, $\alpha' \in (0, 1]$, $\rho \in (0, 1]$, $T_i > 0$ and $\lambda \in (0, 1]$ . Initialize $V_i(r_j)$ randomly for each $i, j$ .

(2) Repeat,

(a) Choose resource $r_j$ according to Boltzmann softmax eq. 3.

(b) Receiving utility $u_j$,

$$V_i(r_j) \leftarrow (1 - \alpha)V_i(r_j) + \alpha u_j.$$

(c) Receiving MGI $u'_k (= U_k(n_k + l))$ where $k \neq j$,

$$V_i(r_k) \leftarrow (1 - \alpha')V_i(r_k) + \alpha' u'_k.$$

(d) If $r_j = \mathrm{argmax}_{r_k} V_i(r_k)$ and $u_j > \rho V_i(r_j)$ or $r_j \neq \mathrm{argmax}_{r_k} V_i(r_k)$ and $u_j < V_i(r_j^*)$,

$$T_i \leftarrow \lambda T_i$$

**Fig. 2.** Procedure of Learning Estimated Utility and Adapting Temperature

learns slowly when it is getting more reward than expected (winning phase), while it learns quickly when it gets worse reward (losing phase). We interpret and abstract this principle as follows:

> An agent should *exploit* in the winning phase, and *explore* in losing phase.

We call this principle as *abstracted WoLF*.

In order to apply the abstracted WoLF, we must define two concepts, *winning* and *exploration/exploitation*. For *winning*, we take the following definition given in [1]

> A player was winning if he'd prefer his current strategy to that of playing some equilibrium against the other player's current strategy.

About *exploration/exploitation*, we simply define that the *exploration* relaxes the condition for an agent to try various possibility, and the *exploitation* forces the agent to examine only neighboring possibility of the current best policy.

## 3.2 Adaptive Temperature Control based on WoLF

As described in section 2, each agent decides its choice according to Boltzmann softmax (eq. 3). Therefore, *exploration/exploitation* can simply be defined as follows:

- *exploration* = to increase the temperature parameter $T$.
- *exploitation* = to decrease the temperature parameter $T$.

We also define *winning* as follows:

(1) Let $\beta \in (0, 1]$ and $\mu > 0$. Initialize $\bar{n}_j \leftarrow 0.0$ and $\hat{n}_j \leftarrow 0.0$ for each $j$ .

(2) In each cycle of agents' decision (step (2) in figure 2),

    (a) Count $n_j$, the number of agents who use resource $r_j$.

    (b) Let

$$\bar{n}_j \leftarrow \frac{\bar{n}_j + \beta n_j}{1 + \beta} \qquad \hat{n}_j \leftarrow \frac{\hat{n}_j + \beta n_j^2}{1 + \beta}$$

$$l \leftarrow 1 + \mu \sqrt{\hat{n}_j - \bar{n}_j^2}$$

    (c) Calculate utilities and MGI

$$u_j = U_j(n_j) \qquad u_j' = U_j(n_j + l)$$

    (d) Provide $u_j$ to the agents who use resource $r_j$, and inform $u_j'$ to other agents.

**Fig. 3.** Procedure of Adapting Moderation Factor

Agent $a_i$ is winning

- when it chooses resource $r_j$ whose expected utility $V_i(r_j)$ is the best in $V_i$, and receives more utility than $V_i(r_j)$, or
- when it chooses non-best resource $r_k$, and receives less utility than $V_i(r_j)$ ($r_j$ is the best resource according to $V_i$).

This definition is reasonable because the learning of the agent will enhance the current best policy ($V_i(r_j)$). As described in section 1, we are focusing the stability of agents' behaviors. Therefore, we take only the best choice of each agent into account, because only the best choice is meaningful in the stabilized situation.

Finally, we get a procedure shown in figure 2.

### 3.3 Adaptive Moderation Control of MGI Method based on WoLF

As mentioned in section 2.2, agents can reach a Nash equilibrium using MGI in learning. However, the convergence to the equilibrium is relatively fragile when the state (agents' estimated utility $V_i$) is far from the equilibrium, because agents tend to behave relatively random in such cases. As a result, MGI brings bad influence to the learning of $V_i$. For example, the top of figure 6 shows the changes of the number of agents who choose a resource in the case the moderation factor $l = 1$. In this case, the learning can not reach the equilibrium so that there remain unstable behaviors of agents. When we choose the moderation factor $l = 5$ (the middle of figure 6), the learning becomes more robust and reach closer to the equilibrium. However, there still remains some perturbation in agent behavior in this result. Generally, larger and larger moderation factor we use, more and more robust the learning is. On the other hand, when we let the large value to the moderation factor, we can not guarantee that the learning can reach Nash equilibrium shown in [7].

In order to attack this problem, we also try to apply the abstracted WoLF principle to the control of the moderation factor. First, we consider RSP as a game between resources and agents: The purpose of the resources is to stabilize agents' behaviors, while the purpose of agents is to get better utilities.

Similar to section 3.2, we define *exploration/exploitation* and *winning* onto the new game. As explained in section 3.1, we consider that the meaning of the *exploration* and *exploitation* are to relax and to tighten player's options, respectively.

Under the context of the new game, we can interpret them as:

- *exploration* = to increase the moderation factor $l$.
- *exploitation* = to decrease the moderation factor $l$.

This interpretation comes from the following consideration: When resources provide MGI using the small moderation factor, the resources suppose that the current situation is close to the equilibrium so that they force agents to exploit neighboring situations. On the other hand, the large moderation factor means to permit for agent to try wider situations.

For *winning*, we simply consider as follows: Because the purpose of the resources is to stabilize the agents' behaviors, we measure the *winning*-ness by changes of the number of agents using each resource. Actually we use the standard deviation of the number of the agents to distinguish *winning* of resources, that is, a resource is winning when the standard deviation of the number of agents is small.

Finally, we get a procedure shown in figure 3 for the control of the moderation factors.

## 4 Experiments

### 4.1 Experimental Setup

In order to show the effect of the proposed method, we conducted several experiments of RSP.

As the utility function, we take the following function:

$$U_j(n) = 1 - \epsilon_j^{\frac{C_j}{n}}, \tag{5}$$

where $\epsilon_j$ and $C_j$ are the error rate and the capacity of the resource $r_j$, respectively. It is obvious that the utility function eq. 5 satisfies the condition eq. 1.

The meaning of this function is as follows: Consider a kind information service in which an agent can get a required information in the probability

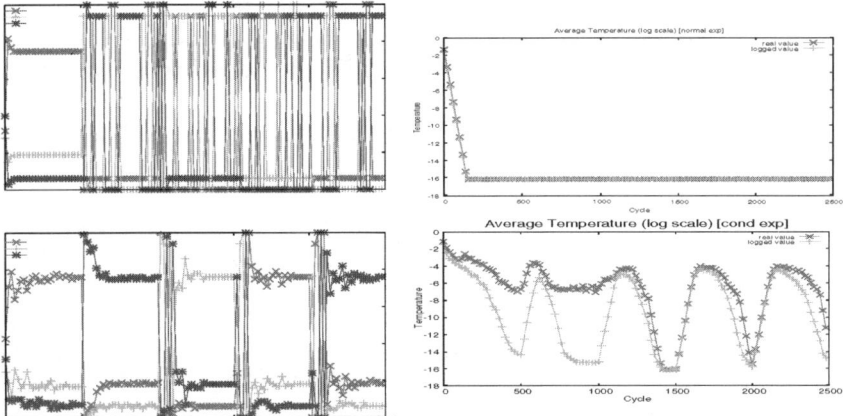

**Fig. 4.** Exp.1: Changes of Number of Agents Who Choose Each Resources in under Adaptive Temperature Control

**Fig. 5.** Exp.1: Changes of Average Temperature under Adaptive Temperature Control

$(1 - \epsilon)$ for a query. The agent can repeat the query $\frac{C_j}{n}$ times in one cycle until it get the required answer. When the capacity $C_j$ is large, the agent can try many times. On the other hand, when the number of agents who use the resource, the agent has few chance to try its query in a certain time-period. Therefore, the above utility function means the probability that an agent can get the right information in one cycle.

In the following experiments, we use the following parameters:

| | |
|---|---|
| the number of agents $n$ | 100 |
| the number of resouce $m$ | 3 |
| the capacities of resource $\{C_j\}$ | $\{20, 5, 2\}$ |
| the error rate $\epsilon$ | 0.1 |
| the learning rate $\alpha$ | 0.1 |
| the learning rate $\alpha'$ | 0.05 |
| the decay factor $\lambda$ | 0.9 |
| the decay factor $\beta$ | 0.9 |
| the margin factor $\rho$ | 0.99 |
| the amplification parameter $\mu$ | 3.0 |

### 4.2 Exp.1: Adaptive Temperature Control

In the first experiment, we evaluate the robustness of the adaptive temperature control proposed in section 3.2 against changes of environments.

As similar to the preliminary experiment shown in section 2.3 and figure 1, we train agents in the environment which changes the capacities of resources $C_j$ in a certain period. For this experiment, we rotate the capacities of resource in every 500 cycles. We use $l = 3$ as the moderation factor for MGI learning. Figure 4 shows the result of changes of the number of agents in each resource through learning. The top graph of this figure shows the case the agent drop the temperature monotonically. As same as the preliminary experiments, while the agents can reach the equilibrium in the first phase (cycle $< 500$), they start oscillating (up-and-down) behaviors at the first change of the environment. After then, the oscillation never stop. On the other hand, when we use the adaptive temperature control based on WoLF (the bottom of figure 4), we can see agents reach an equilibrium in each phase of 500 cycles. We confirmed it by investigating the changes of average temperatures of all agents. The top and the bottom of figure 5 indicate changes of the averages of temperatures of all agents with monotonically dropping and adaptive temperature method in log-scale, respectively. In each graph, both of arithmetic and geometrical means are plotted. From these graphs, we can confirm that agents raise the temperature when the environment is changed so that they start to explore a new equilibrium. When the behaviors get close to the equilibrium, they drop their temperature gradually. As a result, the temperatures become low enough so that the most of agents never change their choice until the environment is changed.

### 4.3 Exp.2: Adaptive Moderation Control

In the second experiment, we shows the convergence performance of adaptive moderation control compared with the fixed value of moderation in the MGI learning.

As mentioned in section 3.3, the learning is fragile with the small moderation factor while the learning can not be guaranteed to reach Nash equilibrium with the large moderation factor. The top and the middle graph in Figure 6 shows the changes of the number of agents using each resources in the case of fixed moderation factor $l = 1$ and $l = 5$, respectively. On the other hand, the bottom shows the result of the adaptive moderation control proposed in section 3.3. Here, we can see that the agents' behaviors are completely converged into the equilibrium. Figure 7 shows the changes of arithmetic and geometrical means of temperatures of all agents in each case. These figures show that agents can not reach equilibrium but continue exploration in the case $l = 1$ (the top), while the most of agents reach equilibrium and stop exploration in the case of $l = 5$ (the middle) and the adaptive moderation. Because the case of the adaptive moderation reaches the stable state completely, the moderation factor $l$ should be 1. This means that the agents' choice is one of Nash equilibrium. Therefore, the state is a kind of optimum so that the average of utilities agents gets is locally maximized. Actually, the final value of the total utilities in the adaptive moderation control is 46.292, which is greater than the other cases (46.239 in $l = 1$ and 46.258 in $l = 5$). Figure 8 shows the changes of average utilities each agent gets through learning. Similar to figure 6, each agent can get stable and relatively high utility in the case of the adaptive moderation control (the bottom), while the utility sometimes drop in the case with of fixed moderation factors because of the perturbation.

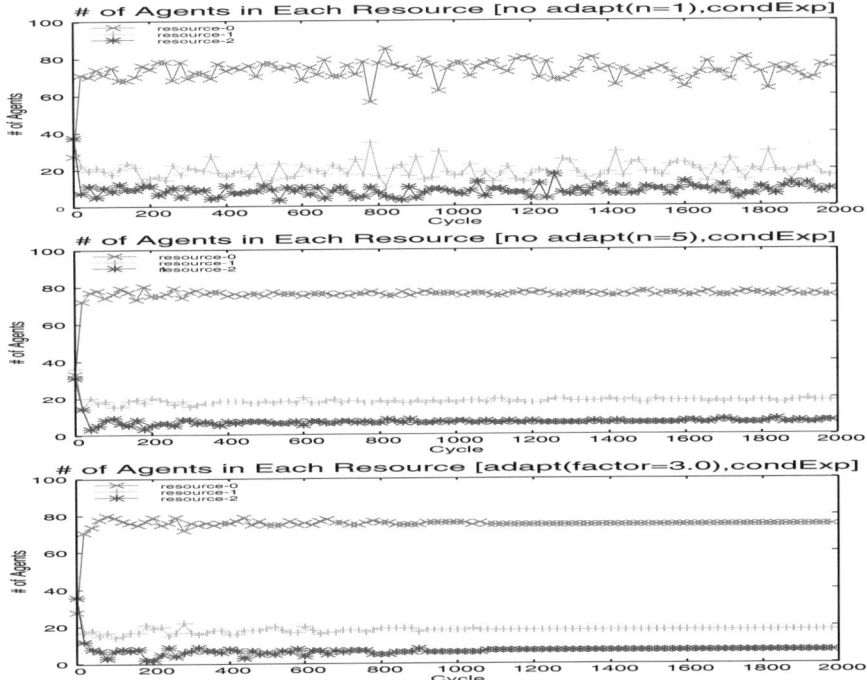

**Fig. 6.** Exp.2: Changes of Number of Agents Who Choose Each Resources in under Adaptive Moderation Control

# 5 Concluding Remark

In this article, we introduce the stability of agents' behaviors as a criterion of the resource sharing problem, and propose methods to control two learning parameters, temperature in Boltzmann softmax function and the moderation factor in the moderated global information, based on the abstracted WoLF principle. The experimental result shows the advantage of the method in the robustness against changes of environment and the steadiness of learning.

Although we show experimental results of a small variation of setting, the proposed methods can be applicable to several environments. For example, we can use different kind of utility functions that satisfy the condition eq. 1. It is also possible to introduce internal states in the agent to adapt Markov decision processes.

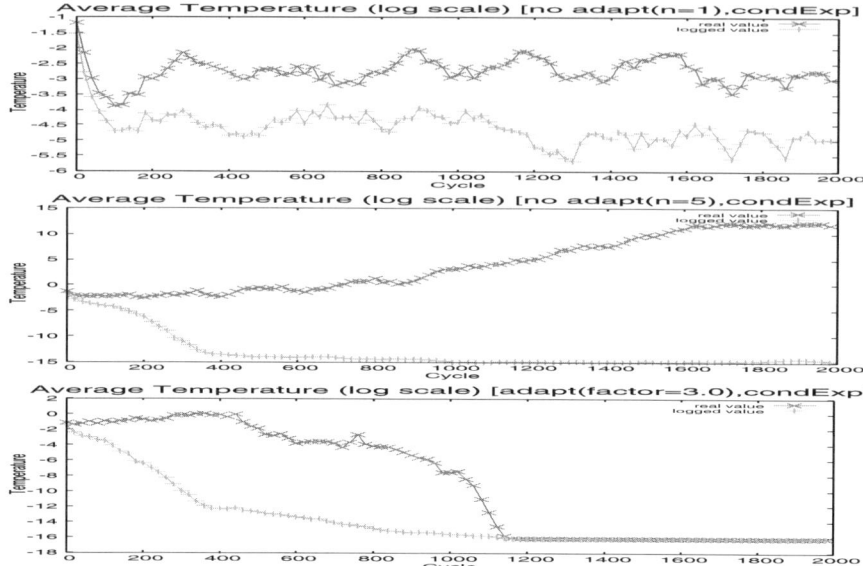

**Fig. 7.** Exp.2: Changes of Average Temperature under Adaptive Temperature Control

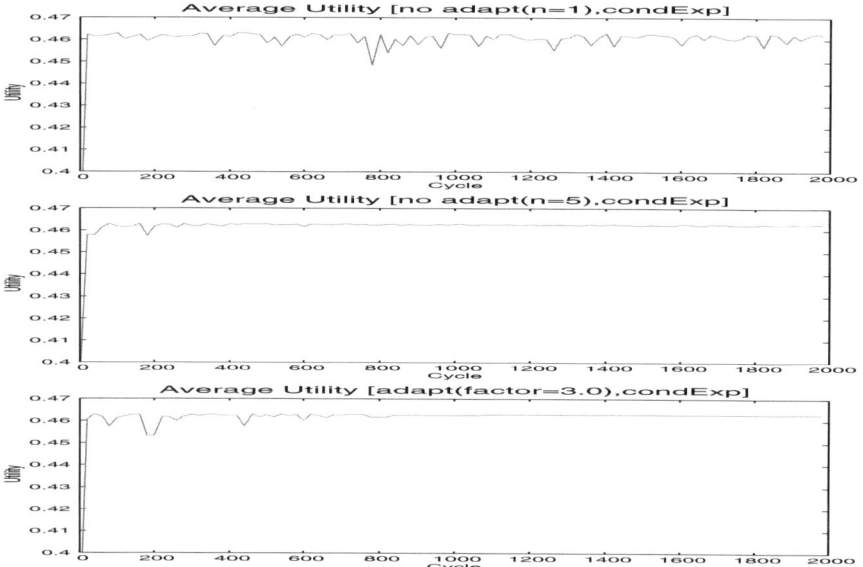

**Fig. 8.** Exp.2: Changes of Average Utility under Adaptive Temperature Control

There remain several open issues on the proposed methods:

- How to set remaining learning parameters suitablly, especially the amplify factor $\mu$ in the adaptive moderation control.
- The relation between the convergence speed and the robustness of the learning and the control.
- How to formalize hieralchical social structures as a resource sharing problem under which groups of agents make choices.

# References

1. Michael Bowling and Manuela Veloso. Multiagent learning using a variable learning rate. *Artificial Intelligence*, 136:215–250, 2002.
2. Michael H. Bowling and Manuela M. Veloso. Rational and convergent learning in stochastic games. In *Proc. of IJCAI 2001*, pages 1021–1026, 2001.
3. David Carmel and Shaul Markovitch. Exploration strategies for model-based learning in multiagent systems. *Autonomous Agents and Multi-agent Systems*, 2:141–172, 1999.
4. Andrew Garland and Richard Alterman. Learning procedural knowledge to better coordinate. In *Proc. of the Seventeenth International Joint Conference on Artificial Intelligence*, pages 1073–1079. IJCAI, 2001.
5. Maozu Guo, Yang Liu, and J. Malec. A new q-learning algorithm based on the metropolis criterion. *IEEE Transactions on Systems, Man and Cybernetics, Part B*, 34(5):2140–2143, Oct. 2004.
6. Franziska Kluegl, Ana L. C. Bazzan, and Joachim Wahle. Selection of information types based on personal utility - a testbed for traffic information markets. In *Proc. of the Second International Joint Conference on Autonomous Agents and Milti Agent Systems*, pages 377–384. AAMAS, 2003.
7. Masayuki Ohta and Itsuki Noda. Reduction of adverse effect of global-information on selfish agents. In Luis Antunes and Keiki Takadama, editors, *Seventh International Workshop on Multi-Agent-Based Simulation (MABS)*, pages 7–16, Hakodate, May 2006. AAMAS-2006. Shinko.
8. Bob Price and Craig Boutilier. Accelerating reinforcement learning through implicit imitation. *Journal of Artificial Intelligence Research*, 19:569–629, 2003.
9. Ming Tan. Multi-agent reinforcement learning: Independent vs. cooperative agents. In *Proc. of the Tenth International Conference on Machine Learning*, pages 330–337, 1993.
10. David H. Wolpert, Kagan Tumer, and Jeremy Frank. Using collective intelligence to route internet traffic. *Advances in Neural Information Processing Systems*, 11:952–958, 1999.
11. Tomohisa Yamashita, Kiyoshi Izumi, Koichi Kurumatani, and Hideyuki Nakashima. Smooth traffic flow with a cooperative car navigation system. In *Proc. of the Fourth International Joint Conference on Autonomous Agents and Milti Agent Systems*, pages 478–485. AAMAS, 2005.

# An Evolutionary Rulebase Based Multi-agents System

Hiroshi Ouchiyama, Runhe Huang, and Jianhua Ma

Faculty of Computer and Information Sciences,
Hosei University,
Koganei-shi, Tokyo 184-8584, Japan
{rhuang, i04t0004, jianhua}@k.hosei.ac.jp

**Summary.** This paper presents an evolutionary rulebase based multi-agents system in which each agent's rulebase is evolving with its increasing experience and knowledge and as well as sharing experience and information about its environment from other agents. The problem of low matching probability between a current environment state and rules in the rulebase that starts from the scratch without prior gained experience and knowledge is addressed and a solution of combining the partial matching with dynamic adaptive matching rate to the problem is proposed. The efficiency of sharing experience and information about the environment among all agents is emphasized and both local sharing and global sharing strategies are described.

## 1  Introduction

A set of rules representing knowledge of an agent for making decisions in the reasoning process has been widely adopted in many intelligent agent systems [2] [3]. There are mainly two approaches. One is using a set of fixed rules, which is quite suitable for a simple, static, and information accessible environments. However, most environments agents are working in are complex, dynamic, and non-accessible. For an agent like exploration agent, a set of evolutionary rules are appropriate. The set of rules can be evolving with increasing agent itself experience and knowledge passed from other agents. The agent experience is implicitly represented or encoded in the strength function for evaluating of rules, the selection scheme for generating new rules using a genetic algorithm, and reflected in the reward scheme that takes into consideration of feedback from the environment.

In a conventional classifier system [1] [13], each rule consists of a condition part and an action part. Both parts are strings out of {0, 1, #}, where the '#' serves as a don't-care symbol. A non-negative real number is associated with each rule indicating its strength. The current environment state is matching with the condition part of all rules in the classifier system. Each matched rule computes and pays a *bid* that is distributed among the rules fired during the last time round. This is so-called the bucket brigade algorithm [14] in which the complete matching of the current state with the condition part of all rules is used. To a rulebase that starts from scratch without prior experience and knowledge, the low matching rate of the current state

H. Ouchiyama et al.: *An Evolutionary Rulebase Based Multi-agents System,* Studies in Computational Intelligence (SCI) **56,** 203–215 (2007)
www.springerlink.com

with the condition part of all rules becomes a problem in terms of the effective and the efficiency of rule evolution. This paper proposes a partial matching scheme instead of the complete matching and the dynamic matching rate is adopted in order to adapt changing environment as time goes.

For most tasks, multi-agents instead of a single agent could get work down quicker and better since the work can be distributed among the agents and they can share what they know about environment or what they learn. This paper takes the object exploration problem to describe how multi-agents collaborate locally and globally. The local collaboration is performed by each agent publishing and sharing some better quality rules. The global collaboration is performed by each agent posting and sharing a virtual memo that contains what the agent think is valuable for other agents. Both local and global sharing strategies are described in details in this paper.

## 2    The Exploration Problem

It is assumed that in a wall surrounded 2-D grid environment where agents are to explore specified items scattered all over. Except for the specified item or items, there may be a rock in a grid, which blocks the exploration path. The goal of each agent is to find the specified items in the environment as many as possible as quicker as possible during a period of time or cyclic rounds.

## 3    Evolutionary Rulebase

### 3.1    The EEERB Overview

The experience and environment based evolutionary rulebase (EEERB) system is the extension of the conventional classifier system. The overview of EEERB system is presented in Figure 1. As any other classifier systems, a set of condition-action representations in EEERB are called rules and a population of rules is stored in the rulebase. The *Meta Rules* describe how to apply the constraints to rules for a specified domain problem and provides the guidance of generating rules to avoid excessive and illegal rules in the rulebase. There are three ways of generating rules: randomly initializing a population of rules, updating the population by adding a certain number of new rules generated using GA and deleting the certain number of worse rules from the population so as to keep the same size of the population, and dynamic generated rules based on public information and facts posted by other agents in the same environment. The *Detector* functions as a translator that inputs the received facts from the environment. The rulebase is scanned and any rule whose condition is matched with the current environment situation at the specified matching rate is put into the current *Matched List* [$ML_t$]. An action is selected by applying the rule selection scheme and all rules in [$ML_t$] that advocate the selected action, are put into the current *Action List* [$AL_t$]. *Effecter* functions as an interpreter that outputs an agent's action on the environment. Reward from the environment is shared and distributed evenly among the rules in [$AL_t$]. The total amount of a fixed fraction of each rule's strength in [$AL_t$] is discounted by a factor and then shared evenly by the rules in [$AL_{t-1}$]. Each individual rule in the rulebase has a strength attached with, which is also used in the genetic algorithm for selecting parent rules to generate offspring. The rule advocated and selected to fire receives a reward. If it receives a positive reward value, and the rule increases its strength. The stronger the rules in the rule base, the more likely they survive, be selected, and make contribution to offspring. Inversely, those not advocated or advocated but not selected receive a negative reward and they decrease their strengths. The rulebase is evolving with the agent itself experience and other agents' experience about the environment. An agent learns not only from itself experience but also from others.

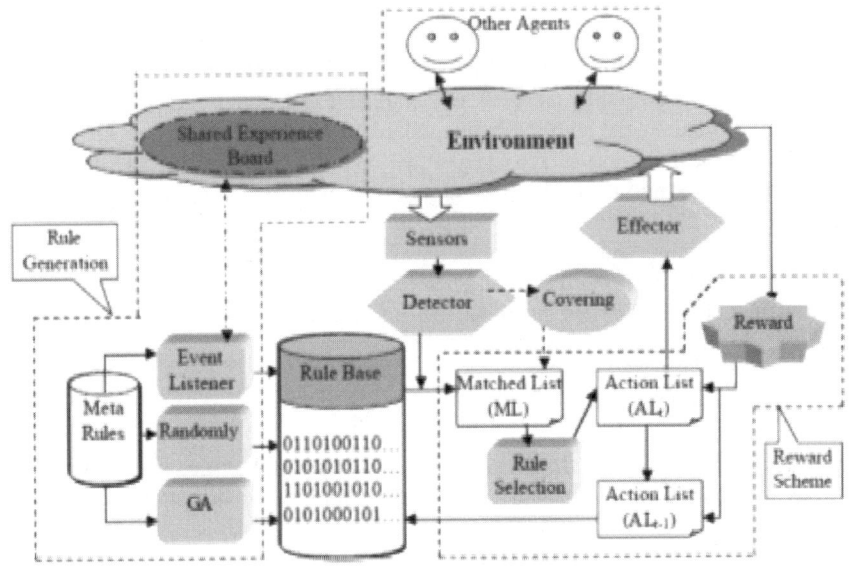

**Fig. 1.** The overview of the EEERB system

## 3.2    Evolutionary Mechanism

For this exploration problem, there are 8 defined actions of going towards 8 neighboring grids for an agent to take and an agent can take an action at a time. 3-bits binary coding strings express the 8 actions. Which action an agent takes depends on the neighboring grids' states which are encoded in 40-bits binary coding strings. An example of the 43-bits coding strings is given below, where '#' acts as a wildcard to express don't care bits.

0##00 0##01 ###10 ###10 ###10 ###10 ###10 11100 : 111

Evolutionary mechanisms have been intensively studies [10] [12]. In this research, genetic algorithm as the evolutionary mechanism is used [5] [6]. The strength value based Roulette wheel algorithm [7] as a selection scheme is used in the GA for selecting parent strings and in the process of selecting rules from the matched list into the action list. Two genetic operators, one-point crossover and mutation are applied to generate new rules.

### 3.3   Experience and Information Sharing

Each agent can only perceive its local conditions such as the states of the grid it is in and the neighboring grids. Of course, it is difficult for an agent to make a better decision only depending upon the limited local information and knowledge. Although perceived information and gained experience about the environment are getting more and wider while it is exploring [4], it takes time. It is effective and efficient to get other agents collaboration by sharing experience and information not only locally and also globally. A rule fired by an agent and evaluated in the environment with a higher strength is posted to the public sharing board to be shared by other agents. It is so-called the local sharing strategy since the condition parts of the rule contains the local situation and the action to take in such situation. However, an action may be locally optimal but not so globally. Taking global information into consideration for an agent is also important. Any agent can draw a virtual memo that contains what the agent know about the environment and place the map in the critical crossing point for other agents. This is so-called the global sharing strategy.

### 3.4   The Reward Scheme

The rules in the action list, $AL_t$ and $AL_{t-1}$ get rewards and the rules in $ML_t$ but not in $AL_t$ pay a certain amount of penalty. Accordingly, their associated strength values are updated. If we denote $S(a, t)$ as the strength of a rule, whose action is $a$ under the condition $M_t$ at time round $t$, then the updated strength of a rule in $ML_t$ but not in $AL_t$ after taking the action $a$ is calculated according to Formula (1).

$$S(a, t+1) = (1-\beta-\tau) S(a, t) \tag{1}$$

And the updated strength of a rule in $ML_t$ and $AL_t$ after taking the action, $a$, is calculated according to Formula (2).

$$S(a, t+1) = (1-\beta) S(a, t) + \beta S(a, t) Reward(a, t) / k \tag{2}$$

Moreover, a rule who was in the previous action list $AL_{t-1}$ can receive a certain amount of reward as well, and its strength value is calculated as Formula (3).

$$S(a, t+1) = S(a, t) + \beta \, \Upsilon S(a, t) Reward(a, t) / k \tag{3}$$

Where, $\beta$, $\tau$, and $\Upsilon$, are the bid cost rate, tax pay rate, and discount factor, respectively. The function, $Reward(a, t)$, returns the reward value for one of the following five possible situations: finding the item, encountering a rock, facing a wall, or entering an empty neighboring grid. The reward values for different situations are differently set. Once they are set, the same set of values is used in all different experiments.

## 4    Matching Mechanism

### 4.1    The Problem of the Complete Matching Mechanism

After detecting a message from the environment, matching rules in the rulebase with the message is performed. As for the exploration problem, an agent in EEERB system has no prior experience and knowledge about the environment, that is to say, the rulebase contains no any rule at the beginning or a few rules generated randomly or received from the public sharing board. The complete matching used in the conventional classifier systems [9] [13] may have a problem of the low matching probability. To the mentioned exploration problem, the matching probability $(P_m)$ can be calculated as follows.

$$P_m = \frac{N_r}{(2^{40} - 1)} \tag{4}$$

Where, $N_r$ is the number of rules in the rulebase. Of course, the use of the wild card '#' can greatly raise the matching probability. For instance, the example string given in Section 3.2 presents a situation in which there is a rock in the east and there are walls in the west and south, which also implies we do not care about the west south, east south and west north as well. For such situation, the matching probability becomes

$$P_m = \frac{N_r}{(2^{21} - 1)} \tag{5}$$

Even so, it still provides a quite low matching probability unless we increase the total number of rules ($>= 2^{21}-1$) in the rulebase. However, it is not practical to increase rules in the rulebase to this size since it would greatly increase the matching and selecting process time.

To solve this problem, this paper proposes the partial matching together with dynamic changing matching rate mechanism.

### 4.2    The Partial Matching Mechanism

A rule consists of the condition part and the action part. The condition part of a rule contains 8 neighboring grid states and each of them uses 5 bits out of 40 bits. The complete matching is to match the detected message with all 8 states in the condition part. The partial matching mechanism is to match a number of states out of 8 states at the defined matching rate from 1 to 8, which is the minimum number of states to be matched.

The partial matching mechanism can improve the matching probability. However, what the most concerned is the quality of the matched rules. To guarantee a reasonable quality, each rule partially matched with the current environment state is inserted into the matched list ($ML_t$) with a priority value attached. The more states are matched, the rule has a higher priority to be selected and inserted into the current action list ($AL_t$). Whether a rule in $ML_t$ is finally selected or not depends on both the matching related priority and the action related strength. The selection scheme adopts the Roulette Wheel algorithm.

It is obvious that with the increasing number of rules in the rulebase, a fixed matching rate is not appropriate. This paper proposes a dynamic matching rate to adapt the continuously updated rulebase and changing environment situation.

### 4.3   The Dynamical Matching Rate with the Covering Technique

The matching rate is initially set as 1. At each time round, the number of the matched rules and the average reward to the rules in the matched list are used to decide the matching rate at the next time round. If the number of the matched rules is over a defined threshold, the matching rate decreases by 1. If the average reward from the environment to the matched rules is lower than a defined threshold, the matching rate increases by 1 but not over 8. Otherwise, the matching rate stays the same. If the matching rate is 1 and the number of the matched rules is still lower than a defined threshold, the covering technique is applied. The covering technique is to generate actions for the detected environment states so as to form rules and insert them into the matched list.

### 4.4   Firing Rules

Each rule fired in the environment before has a strength attached. Based on the priority coefficient and the strength, the Roulette wheel algorithm can decide which rules in the matched list to be selected to the action list. For those rules that were not fired before, there were no the strength values attached. The simulation of applying the rules to the current environment situation is performed. The rule that receives the highest reward from the environment is to be fired. The rules generated by the covering technique have no strength values attached until they are evaluated in the environment. Evaluating rule takes time so it is not very efficient to use the covering technique unless it is really necessary.

## 5   Agent's Collaborations

Each agent on its own can explore only locally and has a limited ability. Without global vision, an agent can not perform globally better. For obtaining the global information, an agent can go through the entire environment but it takes time. For more efficient and effective exploration, collaborations from other agents are indispensable. Having both local and global information and knowledge, an agent is possible to make reasonable decisions. Here, two strategies called the local sharing strategy and the global sharing strategy are introduced respectively.

## 5.1   The Local Sharing Strategy

Each agent posts their rules with the strength over the defined value to the public sharing board. An agent can take any rules from the public sharing board via the *Event Listener* interface and import them to its rulebase as shown in Figure 1. Importing a rule means taking another agent's experience (of making such action under a certain situation) that is implied in the rule's condition part and action part. A shared rule with a higher strength is more likely to be selected to be fired as long as its condition part is matched with the current situation. Having the shared rule, an agent obtains experience and knowledge from others as if it was in the same situation before. Of course, the agent with more experience and knowledge can make a better decision or action. In particular, when an agent has no prior knowledge and experience such as no rule or only a few rules in the rulebase at its infant stage, such local sharing experience in the rule form are even more important and effective. The local information is thus shared among all agents. That is why it is called the local sharing strategy.

## 5.2   The Global Sharing Strategy

The global sharing strategy is another mechanism for agents' collaborations. Each agent draws down what it thinks is important information or experience in a so-called virtual memo for others and places it in a proper crossing point where other agents can get when they pass the crossing point. The memo may contain the locations and the number of the specified item or the map from a point to the location where there is an item or items. In this strategy, agents exchange not only local information and experience but also global information about where they have been around, where there are rocks, and where there are the specified items. They can update the virtual memo while they are making exploration. The virtual memo itself is not shared as rules like the local sharing strategy, but it is placed in a, what it thinks, critical crossing point. In such collaboration way, all agents benefit from each other and quickly get the global image of the environment. With more local and global knowledge and information, agents can perform better and quickly to achieve their tasks.

Figure 2 gives an example of the virtual memo that contains a drawn virtual map (as shown in the left). The locations, {6, 5} and {8, 10} are where the agent $A$ has been before and the black blocks mark the route the agent $A$ went through. The virtual map with a relative coordinates (as shown in the right) is placed at the location {8, 10} using the coordinates {0, 0} instead of {6, 8} and {2, 5} instead of {8, 10}. When another agent such as agent $B$ passes by the location {8, 10}, it can get the virtual map as a reference to help it make a better exploration.

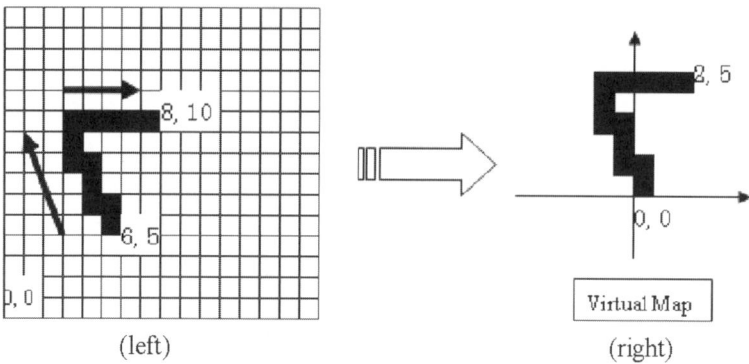

(left)                                              (right)

**Fig. 2.** An example of a virtual map

The global collaboration continues while agents are exploring in the environment. Each agent in such collaboration way can know information about other regions of the environment. That is why it is called the global sharing strategy.

## 6   Implementation and Experiments

Both Java and Jess (Java Expert System Shell [8]) are used as the implementation languages. The rule base and its related reasoning and decision making functions are implemented in Jess. The simulated environment, the genetic algorithm, the selection scheme, the reward scheme, the communication functions, etc. are implemented in Java. A 50x50 2D grid field is set, 4% objects and 10% rocks are generated and randomly placed in the field, and 10 agents are generated and randomly placed in the field. The reward values used in the reward scheme are given as follows.

```
reward(item) = 100
reward(wall) = –10
reward(rock) = 0
reward(rock) = –10
reward(empty) = 10
```

Once the simulated environment is generated, the same simulated environment will be used in all experiments in order to be able to make necessary comparisons under the same environment condition. As for the genetic algorithm, the population size is set 100 rules, the cross over probability is 0.93 and the mutation probability is 0.01 as the most genetic algorithms use.

### 6.1   The Priority Calculation and the Rule Handling

The partial matching mechanism is proposed and its implementation requires the operation on each individual condition of the total 8 conditions in the condition part of a rule. Therefore, a rule in the rulebase has a mirror form stored in 8 split single condition rules as shown in Figure 3. Thus, the matching rate related priority calculation can be easily implemented.

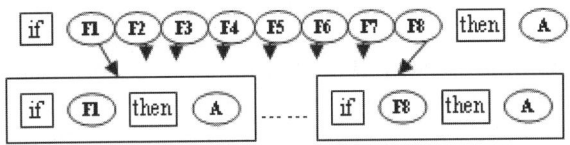

**Fig. 3.** The mirror form of a rule

Jess is used for implementing on rules. A rule in Jess is split into 8 defined rules from c-0 to c-8 as shown in Figure 4.

```
(defrule c-0
 (message
  (north space ?emission&:(or (> ?emission 1) (<= ?emission 3))))
  =>
 (add-matchlist (integer 0)))


 ......

(defrule c-8
 (message
  (north-west space ?emission&:(or (> ?emission 1) (<= ?emission 3))))
  =>
 (add-matchlist (integer 0)))
```

**Fig. 4.** A rule in Jess

## 6.2 The Experiment on the Matching Rate

An experiment is conducted to see how the number of matched different rules is affected by the matching rate. In this experiment, the matching rate is fixed at from 1 to 8, the system runs for 1000 time rounds (cycles) or when the number of different rules in the rulebase reaches 100, and the numbers of different rules in the rulebase at the different matching rate are recorded and given in Table 1.

**Table 1.** The Experiment result on the matching rate

| The matching rate | 1 | 2 | 3 | 4 | 5 | 6 | 7 | 8 |
|---|---|---|---|---|---|---|---|---|
| The number of rules | 26 | 38 | 93 | 100 | 100 | 100 | 100 | 100 |
| The time rounds | - | - | - | 900 | 160 | 150 | 110 | 100 |

As it can be seen that when the matching rate is fixed at between 1 and 3, total number of different rules does not go over 100, it is believe that the agent is moving locally. In contrast, when the matching rate is fixed at between 4 and 8, the number of different rules starts to go over 100 within 1000 time rounds (cycles). It means that the agent is moving around in a wider area within the environment by applying the covering technique. When the matching rate increases, the matching probability decreases as shown in Figure 5. When there is no rule in the rulebase matched with the current situation, the covering technique is applied. The result gives a hint that the matching rate is better to start with a smaller rate and dynamically increasing to a larger rate so as to let the agent to explore from the local to the global in the environment to avoid the problem of local optimization. The matching rate should be adjustable accordingly.

Figure 5 depicts the detailed record of the number of the matched rules in each time round (cycle) up to the maximum 1000 time rounds at the matching rate 3 and 7, respectively. At the matching rate 3, the matched rules at the beginning are only a few but increase as time goes and the rulebase grows. Gradually, the matched rules are increased so that the covering technique is not necessary. While at the matching rate 7, the matching probability is low all the way through. In such situation, the covering technique becomes an important mean for growing the rulebase. It is critical to use an appropriate matching rate together with the covering techniques to achieve a better performance in terms of exploration effectiveness and efficiency. Although we realize and propose to use the dynamic matching rate, the optimization of the matching rate together with the use of covering technique will be one of our future work.

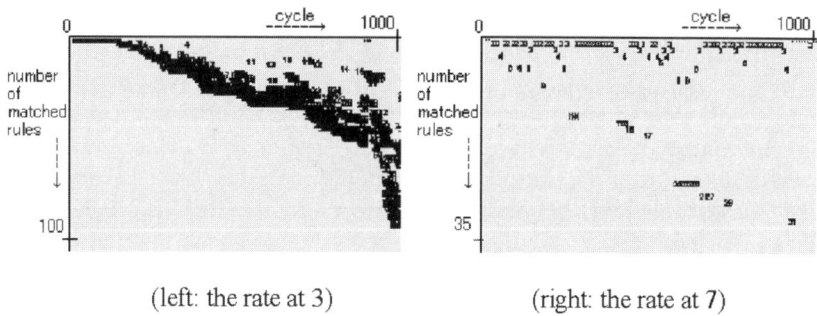

(left: the rate at 3)          (right: the rate at 7)

**Fig. 5.** The number of the matched rules versus the matching rate

### 6.3   Experiments on Different Initial Rulebase

Two experiments that use below two different strategies to generate initial rules in the rulebase are conducted in the same simulated environment.

$s_1$: There is no initial rule in the rule base.

$s_2$: There is an initial set of rules that is generated based on the other agents' experience from the public shared board.

The two experiments are run for 1000 rounds, that is, each agent has 1000 times of interaction with the environment. Each experiment is run for 10 times. The results presented in Table 2 are the average data.

**Table 2.** The results of the two experiments

|  | MNO | ANO | MS | AS |
|---|---|---|---|---|
| $S_1$: No initial rule | 85 | 72 | 223.6 | 170.4 |
| $S_2$:    Have    initial rules | 83 | 71 | 379.7 | 294.3 |

Where, MNO, ANO, MS, and AS denote the maximum number of items found, the average number of items found, the maximum strength of rules in the rule base, and the average strength of rules in the rule base, respectively. The results show that the second initial rule base strategy gives a better performance. It means previous self and others' experience is quite important to an agent. Accumulated experience and knowledge would make significant contribution to build a high quality rulebase of the agent. It also shows that a certain number of population initial rules in the rulebase to start with is better. However, it is necessary to find an optimum population size which will be another future work.

### 6.4    An Agent with Collaboration Versus without Collaboration

Two experiments under the same simulated environment are conducted. In both experiments, the rulebase starts with a population of 100 randomly generated rules. However, an agent in the experiment ($e_1$) works alone to find up to 45 items without any collaboration from other agents in the same environment while an agent in the experiment ($e_2$) works with collaborations from other agents using local sharing and global sharing strategies. Each experiment runs for 10 times and each data given in Table 3 is the average of the 10 runs. Table 3 gives the average time rounds and the maximum time rounds finding an item, respectively. It is obvious that the agent with collaborations from other agents performs significantly better, which is enough to show the important of multi-agents' collaborations.

**Table 3.** The results of the two experiments

|  | Maximum rounds | Average rounds |
|---|---|---|
| $e_1$: Without collaboration | 86 | 30.3 |
| $e_2$: With collaboration | 50 | 20.2 |

### 6.5    The EEERB System Versus the Conventional Classifier System

The same two experiments stated in Section 6.3 are run for the conventional classifier system that uses the complete matching mechanism without local and global sharing strategies. Similarly, each experiment runs for 10 times and the average results are presented in Table 4.

**Table 4.** The performance of the conventional classifier system

|  | MNO | ANO | MS | AS |
|---|---|---|---|---|
| $S_1$: No initial rule | 45 | 34.5 | 148.5579 | 68.81 |
| $S_2$: Have initial rule | 49 | 37.3 | 303.6324 | 151.65 |

Compared the results given in Table 4 with the results given in Table 2, it shows that the EEERB system with partial matching mechanism together with proposed local sharing and global sharing strategies gives better performance in terms of the items found and the rules quality, which proves that it is very important for an agent to learn not only from self experience and also from others.

## 7    Conclusions and Future Work

In this paper, an evolutionary rulebase based multi-agents system is proposed. The system is based on a conventional classifier system but uses a partial matching scheme instead of the complete matching to solve the low matching probability problem. For the partial matching, the dynamic adaptive matching rate is proposed. Moreover, the importance of multi-agents collaboration is emphasized. The local collaboration and global collaboration are explained in this paper. The effectiveness and efficiency of applying the proposed collaboration strategies are presented. The local sharing strategy makes agents to share local knowledge and information while the global sharing strategy makes agents to share global knowledge and information. Comparing with the single agent, multi-agents could get work down quicker and better for most of tasks. Experiments on agents' explorations in a 2-D environment with and without collaborations are conducted. The results show that the agent using collaboration strategies give better performance. Moreover, the experimental results prove that the proposed evolutionary rulebase based multi-agents system performs better than the conventional classifier system in terms of the number of items found in a limited time rounds and the quality of the rulebase.

There are still many remaining work. However, our future work will be focused on finding out

- how many initial rules are necessary in the rulebase so as to achieve an optimal performance, and
- what is a better combination of the partial matching rate and the covering technique.

It is also urgent to develop a dynamic rewarding mechanism since the real world where agents are is changing as time goes. The current system evaluates rules using the fixed rewards. However it will obtain better performance of dealing with real world problems if the dynamic rewarding mechanism is used.

# References

1.  Holland, J. H. and Reitman, J. S.: *Cognitive Systems based on Adaptive Algorithms*, In Waterman and Hayes-Roth, pages 313-329, 1987.
2.  Charniak, E. and McDermott, E.: *Introduction to Artificial Intelligence*, Addison-Wesley Reading, MA, 1985.
3.  Waterman, D.A.: *A Guide to Expert Systems*, Addison-Wesley Publishing Company, Reading, MA., 1986.
4.  Jürgen Schmidhuber: *A Local Learning Algorithm for Dynamic Feedforward and Recurrent Networks*, http://www.idsia.ch/~juergen/bucketbrigade/.
5.  Sakawa, M., Tanaka, M.: *Genetic Algorithm,* Asakura-Shoten, 1995.
6.  David Goldberg, *Genetic Algorithms in Search, Optimization and Machine Learning*, Addison-Wesley, 1989.
7.  IBA Laboratory, The University of Tokyo http://www.iba.k.u-tokyo.ac.jp/english/GA.htm.
8.  Ernest Friedman-Hill: *Jess in Action*, Manning, 2003.
9.  Steward W. Wilson, *ZCS: A Zeroth Level Classifier System,* Evolutionary Computation 2(1):1-18, 1994.
10. Didier Keymeulen, Kenji Konaka, Masaya Iwata, Yasuo Kuniyoshi, and Tetsuya Higuchi. *Robot learning using gate-level evolvable hardware*. In The Sixth European Workshop on Learning Robots, page 173, Brighton, UK, 1998.
11. Yuchao Zhou, *An Area Exploration Strategy Evolved by Genetic Algorithm*, the Graduate Faculty of the University of Georgia in Partial Fulfillment of the Requirements for the Degree "Master of Science", 2005.
12. Franziska Klugl, *Simulated Ant Colonies as a Framework for Evolutionary Models*, In Proc. Of the International Conference on Intelligent Methods in Processing and manufacturing of Material, Vancouver, 2001.
13. Holland, J. H.: *Properties of the bucket brigade*. In *Proceedings of an International Conference on Genetic Algorithms*. Iillsdalc, NJ., 1985.
14. Juergen Schmidhuber, Classifier Systems and the Bucket Brigade http://www.idsia.ch/~juergen/bucketbrigade/node2.html, 2003-02-21.

# Improvements in Performance of Large-Scale Multi-Agent Systems Based on the Adaptive/Non-Adaptive Agent Selection

Toshiharu Sugawara[1], Kensuke Fukuda[2], Toshio Hirotsu[3], Shin-ya Sato[4] and Satoshi Kurihara[5]

[1] NTT Communication Science Laboratories, Kanagawa 243-0198, Japan
   `sugawara@entia.org`
[2] National Institute of Informatics, Tokyo 101-8430, Japan
   `kensuke@nii.ac.jp`
[3] Toyohashi University of Technology, Aichi 441-8580, Japan
   `hirotsu@ics.tut.ac.jp`
[4] NTT Network Innovation Laboratories, Tokyo 180-8585, Japan
   `sato@ingrid.org`
[5] Osaka University, Osaka 567-0047, Japan
   `kurihara@sanken.osaka-u.ac.jp`

**Summary.** An intelligent agent in a multi-agent system (MAS) often has to select appropriate agents to assign tasks that cannot be executed locally. These collaborating agents are usually determined based on their skills, abilities, and specialties. However, a more efficient agent is preferable if multiple candidate agents still remain. This efficiency is affected by agents' workloads and CPU performance as well as the available communication bandwidth. Unfortunately, as no agent in an open environment such as the Internet can obtain these data from any of the other agents, this selection must be done according to the available local information about the other known agents. However, this information is limited and usually uncertain. Agents' states may also change over time, so the selection strategy must be adaptive to some extent. We investigated how the overall performance of MAS would change under adaptive strategies. We particularly focused on mutual interference by selection in different workloads, that is, underloaded, near-critial and overloaded stituations. This paper presents the simulation results and shows the overall performance of MAS highly depends on the workloads. Then we explain how adaptive strategies degrade overall performance when agents' workloads are near the limit of theoretical total capabilities.

## 1 Introduction

An intelligent agent in a multi-agent system (MAS) often has to select other appropriate agents to assign tasks to perform those that are locally non-executable

T. Sugawara et al.: *Improvements in Performance of Large-Scale Multi-Agent Systems Based on the Adaptive/Non-Adaptive Agent Selection*, System Studies in Computational Intelligence (SCI) **56**, 217–230 (2007)
www.springerlink.com

or to improve the system's load-balancing or efficiency. These partner agents for collaboration are usually determined based on their skills, abilities, and specialties, which are required to execute tasks with the required quality. However, if multiple candidate agents still remain, a more efficient agent is preferred. This efficiency is generally affected by the agents and their environmental states, such as their workload, communication bandwidth, and traffic, including the frequency of task requests, as well as their intrinsic capabilities, such as CPU power.

In an open systems environment like the Internet, however, complete information about the entire MAS is unavailable because it is vast and constantly changing, which means that it is impossible to understand the global states of an open system and the total states of all agents working there. Therefore, agents have to learn and select which agent is more appropriate (i.e., efficient) to collaborate with only on the basis of locally available information. This information is, however, limited and often uncertain.

The aim of this research was to investigate how the overall performance of the total MAS would be affected when all agents selected partner agents autonomously based on their subjective views and local partner selection strategies (PSSs) through observation and learning, where 'overall' does not means the performance of individual agents but the average performance of all agents. This issue is important in designing better PSSs for all agents. The overall performance is influenced by the environment and agent states as well as PSSs. Furthermore, we have to consider interference by selection. For example, a high performance agent, $a$, is identified as the best partner by many agents and is thus likely to be overloaded. As this concentration worsens the observed data on performance, many agents switch to other lower-performance agents. Later, $a$ is again observed as a high-performance agent, so they switch to it again. This situation becomes complicated if thousands of agents interfere with one another. Of course, the degree of mutual intereference may depends on the workloads of the system. Because of the dynamics of mutual interference, it is important to introduce adaptability into PSS and investigate better learning parameters related to the exploration-or-exploitation dilemma.

In this paper, we discuss how this interference affects overall performance though adaptive or non-adaptive learning with some variations in the learning parameter values at the difference levels of workloads. To measure overall MAS performance, we used the response time from the time a task is sent to when the result is returned, and the dropped number of tasks because the server was overloaded. We will now discuss how the values of these performance parameters change when autonomous partners are selected locally. We already reported some limited cases as a preliminary experimental result in [8]. In this paper, we will illustrate additional experiments with further discussion to understand some observed phenomena which are against our intuition. This paper is organized as follows. The next section explains the model and settings in our multi-agent simulation and how agents learned by using the observed data. We then present the experimental results for three cases in which agents only used locally available information with different workloads. In particular, we compare these results with those in Scharef et al. [6] because their simulation settings were similar to ours. Finally, we discuss some implications and further analyses of "adaptability versus inadaptability" and "efficiency versus fewer task drops."

## 2 Simulation and Model

### 2.1 Simulation Model

The objective of our simulation was to understand the overall performance of a complete MAS in which rational agents selected their partner agents based on local information and how overall performance could be improved by adaptive or non-adaptive learning of PSSs. Before describing these learning methods, we will explain the simulation model for our experiments, which was derived from various Internet services (such as grid computing), which is one of important applications of MAS.

Our simulation model consists of a set of agents, $A = \{a_i\}$, and server agents, $S = \{s_j\}(\subset A)$, that can execute a specific task, $T$. When agent $a_i$ has $T$, it assigns it to a server that it knows, $S_i(\subset S)$. A PSS corresponds to a method whereby $a_i$ selects $s_j$ from $S_i$. To understand the relationship between overall performance and PSSs more clearly, we simplified the other parameters and assumed that all tasks were only of a single type. Note that in this paper a server denotes an agent that can execute a specific task. The set of servers is static because we have only considered a single type of task. Multiple tasks generally request one another from the agents in an MAS, and different agents become servers for other tasks. We did not consider a simple client-server model but rather more flexible multi-agent models in which each agent selects appropriate partner agents depending on the types of tasks from the local viewpoint. Therefore, agents in an MAS have to select appropriate partners, task by task; our simulation model corresponds to a single attribute in these selections for a certain type of task.

One observable parameter concerning agent performance on the actual Internet is response time ($rt$), i.e., the length of time between when an agent requests a task and when the result is returned. We assumed that a smaller $rt$ would better than a larger one. Another parameter for evaluating agents is the number of dropped tasks, which indicates that the requested task has been silently discarded or refused with a notice, which is also better if smaller numbers are used. Other parameters such as the type of CPU and the number of queuing tasks in servers can also be used to evaluate agents; however, since these parameter values are not usually locally available, they were not used by agents to evaluate servers in our experiments.

Hence, using the observed data, each agent learns which server's $rt$ will be the smallest. Agent $a_i$ in our experiments calculated expected response time $e_i^j$ of known server $s_j$ and (usually) selected the best server, $arg\ min_{s_j \in S_i} e_i^j$.

We adopted the average value for the observed response time in all agents to evaluate the overall performance of an MAS. This value was denoted by $RT$. Thus, our experiments demonstrated how $RT$ changes when all agents attempt to make $rt$ smaller. Another parameter enabling the total MAS to be understood is drop count $D$, which is the total number of dropped tasks. Note that a smaller number of dropped tasks is better, and drops are only observed when the MAS is overloaded.

## 2.2 Simulation Environment

In our simulation, $|A| = 10000$ and $|S| = 120$. All agents were randomly placed on the points of a 150 x 150 grid plane with a torus topology. We also introduced a Manhattan distance to this grid. We assumed that one tick would correspond to 10 ms in this simulation setting . For every tick, $tl$ tasks are generated and passed to $tl$ randomly selected agents, where $tl$ is a positive integer called the *task load*. This is denoted by $tl$ task/tick or simply $tl$ T/t. After receiving a task, each agent $a_i$ immediately selects server $s_j \in S_i$ using its PSS and sends the task to $s_j$. Server $s_j$ processes the committed task and returns the result to $a_i$. Agent $a_i$ can observe the response time for every task and calculate expected response time $e_i^j$ that will be used in the next PSS. The response time is the sum of the durations for communicating, queuing, and processing.

All servers were assumed to have their own CPU capabilities, i.e., each server could process a task in 100 to 500 ms. All servers had different capabilities randomly assigned *a priori*. These capabilities are invariant once they are assigned. When a task arrives at server $s_j$, it is immediately executed if $s_j$ has no other tasks. The result is then returned. If $s_j$ has other tasks, the received task is stored in its local queue and queued tasks are processed in turn. An agent can store 20 tasks in its queue. If $a_i$ already has 20 tasks in its queue, the new task is dropped or refused.

The communication cost, which is the time required to send a task, is assumed to be proportional to the distance between the agent and server and ranges from 10 to 120 ms (we assumed the nation-wide area of the Internet). Of course, the communication cost is also a parameter of the simulation environment, although it was not altered in any of the experiments discussed in this paper, so that we could more easily understand the effect of PSSs based on local observation. All agents have their own scope depending on the distance (less than 14 in our experiments), so they can initially recognize and communicate with all agents and servers within their scope. For all agents to compare server response times and to select the best (or a better) one, they have to identify at least two servers. If an agent does not know any or only one server, it asks all known agents for known servers; by doing this, agents could initially recognize two to fifteen servers in the experiments.

Parameters $RT$ and $D$ in our experimental performance data were first calculated every 20-K ticks. These data improved over time by agents learning. We assumed that $RT$ and $D$ after learning would be the average values of those that were observed during 600-K to 800-K ticks (so each value was the average for ten data), since these data do not vary much after 600 K. The average values for three independent experiments derived from three series of random numbers using three different seeds were then utilized as the experimental results for our simulations. Each of the total sums of the capabilities by all agents was theoretically between 4.7 and 5.0 tasks per tick in these three experiments. We evaluated the performance data when $tl = 1, 4$, and 5. The $tl = 1$ corresponded to underloaded and the $tl = 5$ to overloaded situations. The $tl = 4$ was under but near the limit of their theoretical total CPU capabilities, so is assumed to be the near-ciritical situation.

## 2.3 Estimates of Response Time

As agents select appropriate servers using expected response time $e_i^j$, the calculation of these values is an important part of PSSs. We used the *average values* of observed response time $h_i^j$ and the values estimated by *update function* $w_i^j$, which are often used in reinforcement learning, in our experiments. They are calculated with the following formulas:

$$h_i^j[n] = h_i^j[n-1] * (1 - 1/n) + rt_i^j[n] * 1/n \qquad (1)$$

$$w_i^j[n] = \begin{cases} w_i^j[n-1] * (1 - \lambda) + rt_i^j[n] * \lambda & \text{(if } n > 1) \\ rt_i^j[1] & \text{(if } n = 1), \end{cases} \qquad (2)$$

where $rt_i^j[n]$ is the $n$-th observed response time when agent $a_i$ sends the task to server $s_j$ and $\lambda$ is the learning parameter ($0 \leq \lambda \leq 1$, $\lambda = 0.2$ in the experiments discussed in this paper). Value $h_i^j[n]$ (or $w_i^j[n]$) is $a_i$'s expected response time to server $s_j$ obtained by Eq. 1 (or Eq. 2) after $n$ response-time data, $r_i^j[1], \ldots, r_i^j[n]$, for $s_j$ were observed. We can describe $h_i^j$ and $w_i^j$ simply if value $n$ is unnecessary. Note that, $\lim_{n \to \infty} 1/n = 0$, so $h_i^j$ becomes stable, although $w_i^j$ may adaptively change according to the performance of the server. After this, Eq. 1 will be called the *average value function*.

# 3 Experimental Results and Analysis of Performance

In the first experiment (Exp. 1), agent $a_i$ selects one server from $S_i$ every 500 ticks using the following PSS.

P1. Agent $a_i$ selects server $arg \min_{s_j \in S_i} e_i^j$ with a probability of $p$ ($0 \leq p \leq 1$). If multiple servers have the best $e_i^j$, one is randomly selected.

P2. Otherwise, (as with probability $1 - p$), $a_i$ selects the server with probabilistic distribution $Pr(s_j)$,

$$Pr(s_j) = (e_i^j)^{-l} / \sum_{s_k \in S_i} (e_i^k)^{-l} \qquad (3)$$

where $e_i^j = h_i^j$ or $w_i^j$. Agent $a_i$ initially sets $e_i^k = 0$ for known server $s_k$ so that it first selects $s_k$ with no observed data.

Probability $p$ is a kind of exploration-or-exploitation parameter in the sense that $p$ controls whether an agent exploits the learning result (so it selects a server whose expected response time is the best at the time) or explores another server's performance (so it can select a server that is not the best). Power $l$ is also a kind of exploration-or-exploitation parameter and introduces some *fluctuations* since an agent may select a server whose $e_i^j$ is not the smallest. Note that if $l$ is larger, these fluctuations are small but $p$ and $l$ have different effects on PSSs. Suppose that an agent knows two servers and their observed response times are 100 and 101 ticks. The server with a response time of 100 ticks is obviously the best. If $p = 0.9$, then the best server is selected with this probability. However, the best server is selected with probability 0.512 calculated with Eq. 3 even if $n = 5$. This probability decreases if it knows more servers.

■ Using update function ($p$=0.9) ○ Using update function ($p$=0.8) ◇ Using average value function ($p$=0.9)

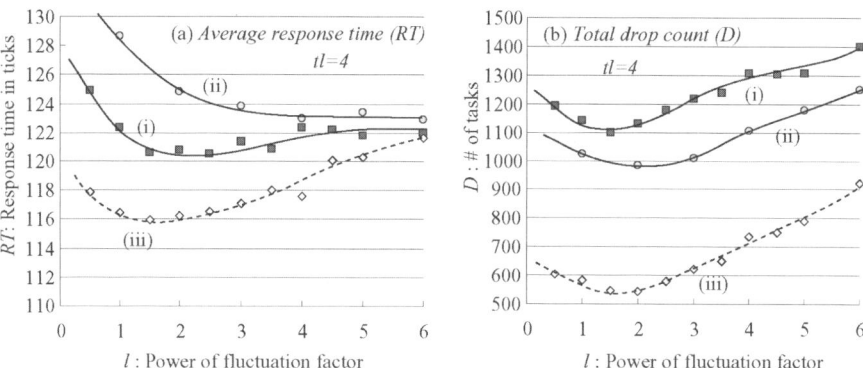

**Fig. 1.** Overall performance values after learning in near-critical situation ($p = 0.9$ and 0.8)

First, we assumed $tl = 4$ (T/t), which is the near-critical situation. The experimental data in Figs. 1 and 2 illustrate how independent learning by individual agents can uprate the overall performance of the entire MAS, probably by load-balancing; actually, since the first observed $RT$ and $D$ values, i.e., during 0 K to 20 K ticks, are 272.9 ticks and 11180.0 tasks, respectively, they can be considerably improved.

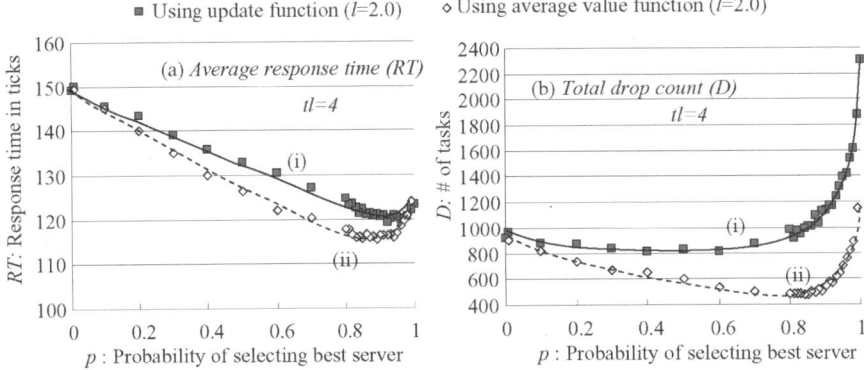

**Fig. 2.** Overall performance values after learning in near-critical situation ($l = 2.0$)

Figure 1(a) indicates that $RT$ generally improves if $l$ is larger only when $p = 0.8$ when using update function ($e_i^j = w_i^j$). In other cases, especially when using average value function ($e_i^j = h_i^j$), $RT$ is optimal around $1.5 \leq l \leq 2$. This means that some fluctuations can improve overall performance. This is not consistent with the results for the fixed load case obtained by Scharef et al. [6]. An agent's rational decision based on locally calculated value $w_i^j$ will not always lead to better results in a large-scale MAS. This also suggests that using the average value functions makes agent server selections stable and conservative over time, but the convergent state is not optimal.

The changes in drop count $D$, as plotted in Fig. 1 (b), express similar shapes in both cases; a larger $l$ produced many dropped tasks and $D$ is minimal at around $l = 1.5$ to 2 counter to what we had intuited. When agents were using update functions, we could observe a tradeoff between $RT$ and $D$ by comparing (i) and (ii) in Fig. 1; when $p = 0.9$, because all agents were likely to select the best server, many tasks were sent to a few high-performance servers. However, this increased dropped tasks. When $p = 0.8$, fewer dropped counts were observed but their $RT$ values were slightly worse.

Therefore, we investigate how $p$ affected overall performance by fixing $l$ to 2.0 since $RT$ and $D$ were optimal around $p = 1.5$ to 2.0 from Fig. 1. These results are plotted in Fig. 2. The $RT$ values are minimal at around $p = 0.85$ to 0.9 in both cases, and their $D$ values are minimal near 0.6 (when using the update function) and 0.85 (when using the average value function). These graphs show that the observed tradeoff in the previous experiment is a special phenomenon; when $p \leq 0.8$, agents select the best servers by learning using information gathered from their own circumstances, so their $RT$ and $D$ performance values indicate improvement when $p$ becomes larger. However if $p > 0.9$, this PSS induces task concentration to a few servers that have high

CPU capabilities.[6]  Consequently, both $RT$ and $D$ degrade. This simulation revealed that there is a tradeoff when $0.8 \leq p \leq 0.9$, which is a kind of meso-morphic state.

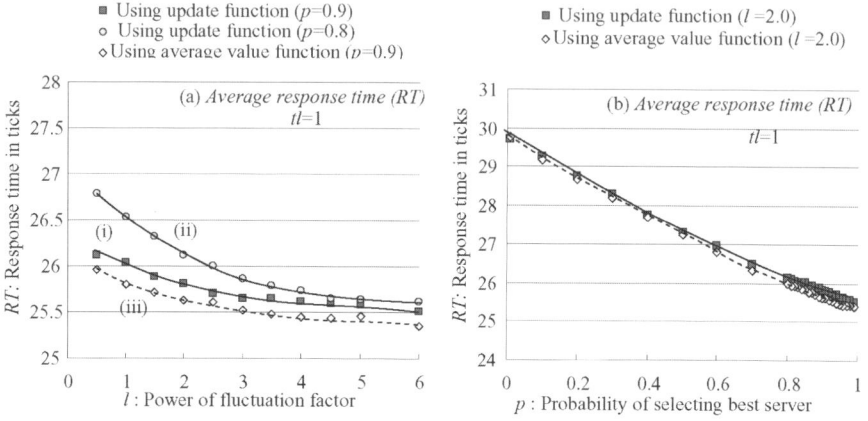

**Fig. 3.** Overall performance values after learning in underloaded stituation ($p = 0.9$, and 0.8)

The most noticeable feature of these graphs is that using the average-value function outperforms the update functions in both $RT$ and $D$. When agents use update functions, they can act more adaptively with the environment including other agents. Hence, this adaptability also continuously changes the environment. Agents can only detect these changes by observing a number of worse response-time data (often many tasks are dropped at this moment) later and this might cause instability; therefore, the overall performance does not improve. However, using average values, agent selections become stable and conservative. Of course, this convergent state is not optimal because their performance is actually improved by introducing a few fluctuations to their decisions; however, it is better than the unstable situation from the viewpoint of overall performance. More discussion on this will appear in the next section.

When $tl = 1$ as in Fig. 3, we cannot observe clear differences between the results for average value and update functions. Predictably, $RT$ improves when $l$ and $p$ are larger. Because the $tl$ is small, the concentration scarcely lead to overload. Note that no dropped tasks can be observed when $tl = 1$. Similarly, when $tl = 5$, the overall performance was improved if $p$ and $l$ were larger as can be seen in Fig. 4. This can also be explained as follows; task allocation

---

[6] Another possible explanation is that agents could not learn all known servers enough because $p$ was large. However, as the experimental data when $tl = 1$ and 5 described below, we believe that all agents could sufficiently learn after 600-K ticks.

■ Using update function (*p*=0.9) ○ Using update function (*p*=0.8) ◇ Using average value function (*p*=0.9)

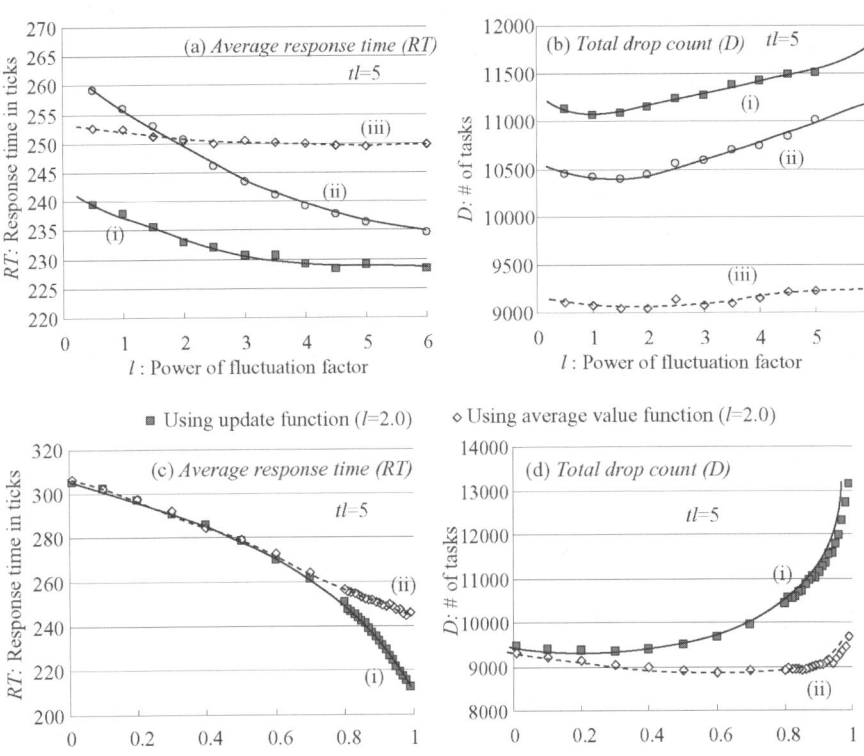

**Fig. 4.** Overall performance values after learning in overloaded stituation ($p = 0.9$, and $l = 2.0$)

has also been concentrated on high-performance agents in this case, but all other agents are already busy because of their heavy task load. As a result, if $p$ or $l$ were larger their average response times would have improved. Instead, the drop count increased. This feature when $tl = 1$ and 5 is consistent with Scharef et al.'s results [6], although it is different from their results when $tl = 4$. Therefore, we can surmise that their fixed load [6] corresponds to a busy or an unbusy case. However, our experiment suggests that the task load has an influence on this feature.

Comparing $RT$ and $D$ when $tl = 4$ and 5 (Figs. 1, 2 and 4), we can find another interesting characteristic. When $tl = 4$ (Figs. 1 and 2), both $RT$ and $D$ are improved together. For example, in Fig. 1, both $RT$ and $D$ worsen when $n$ becomes larger ($n \leq 2$). This means that the structure for agent selection is well-organized at around $n = 2$ and can thus reduce both values. However, $RT$ and $D$ have opposite tendencies when $tl = 5$, i.e., $RT$ improves but $D$ worsens when $l$ or $p$ becomes larger. When $tl = 5$, the number of tasks exceeds the capabilities of all agents; therefore, we can conclude that $RT$ improves with the sacrifice of many dropped tasks. Therefore, MAS performance has quite different characteristics depending on whether task load is low, near a critical point, or overloaded.

In the overloaded situation, the use of the update function can reduces $RT$ (so tasks are executed efficiently). Because agents have many tasks, their average function becomes stable (i.e., less learning effect) at an earlier stage. Therefore, agents cannot react to deviations in the appearance of tasks, but $D$ is smaller, instead. This topic will be discussed in more detail below.

# 4 Discussion and Related Research

## 4.1 Adaptability or Inadaptability

Many task allocation methods have been proposed as mentioned later. Our experiments revealed that controlling task allocation is not obvious when the number of agents is large and each agent has sophisticated capabilities such as decision-making using its own information, learning, or collaboration/coordination. As all the figures in our experiments when $tl = 4$ or $5$ clearly indicate that adaptation using the update function makes the environment fluid, some degradation can be observed, and especially in the near-critical situation, both $RT$ and $D$ are worse when agents use the update function. We assumed quite simple situations, i.e., tasks would be given at a constant rate because the objective of these experiments was to investigate how task load and learning parameters affected overall performance. Nevertheless, the performance of all servers is dynamically changed by the wobble in task distribution. To understand this phenomenon, we examined how many times agents changed their decisions about the best servers that had been determined by $arg$ $\min_{s_j \in S_i} e_i^j$ every 500 ticks over time (this value will be denoted by $C_{bs}$ after this). The results are plotted in Fig. 5 (a), where we observed $C_{bs}$ per 20-K ticks (as well as $RT$ and $D$). The first column in Table 1 lists the average values for $C_{bs}$ from 600-K to 800-K ticks. Note that an agent has 40 ($= 20K/500$) chances per 20-K ticks to change its mind on the best server. Therefore, they can change their decisions on the best server a maximum of 395200 ($= 40 * 9880$) times. Fig. 5 (a) indicates that their decisions about the best server are labile when task load is heavy.

The experimental data listed in the first column of Table 1 can also be viewed as how often the structure of the agent network to determine the best server varied after learning, i.e., the heavier the task load, the more changeable the structure. This was identified by all agents as the environment changes, especially the change in server performance. Furthermore, we can see from Table 1 that the structure is inclined to change more when agents use the update function for their learning. Therefore, adaptability using the update

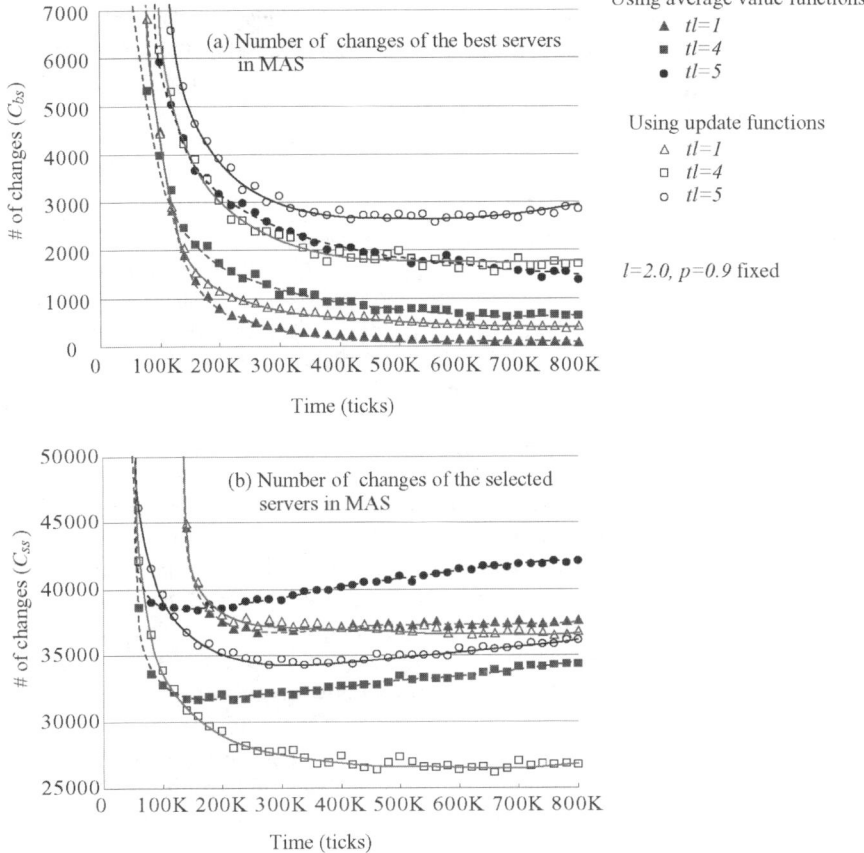

Using average value functions
▲ *tl=1*
■ *tl=4*
● *tl=5*

Using update functions
△ *tl=1*
□ *tl=4*
○ *tl=5*

*l=2.0, p=0.9* fixed

**Fig. 5.** Changes in best servers and selected servers in all agents over time

**Table 1.** Changes in best servers and selected servers in all agents

| Learning function | Using average value function | | | Using update function | | |
|---|---|---|---|---|---|---|
| Task load | 1 | 4 | 5 | 1 | 4 | 5 |
| # of changes in best server decision ($C_{bs}$) | 112.30 | 646.07 | 1572.27 | 428.2.30 | 1691.40 | 2763.43 |
| # of changes in server selection ($C_{ss}$) | 37486.2 | 33999.8 | 41812.2 | 36743.4 | 26655.1 | 35752.2 |

$p = 0.9$ and $l = 2.0$ are fixed.

function also leads to instability, which results in uncertainty about the agents' decisions.

These data also suggest that there is a trade-off between the speed of environmental changes versus the speed of adaptation. To check this, we measured

**Table 2.** Performance values after learning when $\lambda$ is changed

| $\lambda$ | 0.1 | 0.2 | 0.3 |
|---|---|---|---|
| Response time ($RT$) | 123.64 | 120.76 | 121.68 |
| Drop count ($D$) | 1154.8 | 1130.7 | 1227.8 |

$p = 0.9$ and $l = 2.0$ are fixed.

the $RT$ and $D$ when $\lambda = 0.1$ and $= 0.3$ ($p = 0.9$ and $l = 2.0$ were fixed) in a preliminary experiment. The results are listed in Table 2. Both $RT$ and $D$ are optimal when $\lambda = 0.2$. Here, we can identify another tradeoff; when $\lambda = 0.1$, all the agents are a little more stable but adaptation speed is low; however, when $\lambda = 0.3$ they are more aggressively reorganized.

### 4.2 Response Time Versus Drop Count

We can assume that all servers are overloaded when $tl = 5$ because the task load is beyond the theoretical capability of all servers. Therefore, $RT$ improves at the sacrifice of many dropped tasks or *vice versa*. If we compare the graphs in Figs. 1, 2, 3, and 4, it is puzzling why the drop count is always lower when using the average function, or why the response times using the update function improve more than the ones using the average function if the task load is high. To understand this better, we counted how many times agents changed their server selections (this value is denoted by $C_{ss}$), which was derived from P1 and P2. The results are plotted in Fig. 5 (b). The average values for $C_{ss}$ from 600-K to 800-K ticks are also listed in the second column of Table 1.

When $tl = 1$, there is no obvious distinction between using update and average value functions. Interestingly, when $tl = 4$, we can observe a phenomenon where $C_{bs}$ is large but $C_{ss}$ is small if agents use the update function. Fluctuations are mainly introduced into PSS by the P2 discussed in Section 3 in which Eq. 3 is used with a probability of 10%. As this phenomenon indicates that the best server for all agents can still be selected with a high degree of probability with Eq. 3, the differences between the expected response times for the best and second best servers in agents are larger than those in the other cases. This leads to a concentration of high-performance servers thus causing many drops and a biased overload. This worsens both $RT$ and $D$ when using the update function when $tl = 4$. However, when $tl = 5$, all servers are overloaded. The task concentration when using the update function results in shorter $RT$ at the sacrifice of a larger $D$. However, why $C_{ss}$ becomes lower when $tl = 4$ is not known so further study is required to find the mechanism responsible.

Some characteristics always change in actual Internet environments. Hence, we believe there is the tradeoff between unstable and stable; in other words, adaptable and inadaptable. [5] showed that task arrivals in most Internet services are characterized by a heavily-tailed distribution rather than being random or constant, which yields to a bursty nature, so changes are occasional but rapid and drastic, then quickly disappear. Hence, we believe that further simulation and analysis are required to investigate if adaptability or inadaptability is better in such bursty traffic.

### 4.3 Related Research

There have been a number of related research projects on these topics. [4] showed that a learning automaton in a distributed processing environment could improve the load balancing between processors. In the AI field, reinforcement learning was applied to load balancing for distributed/multi-agent systems (e.g., [6, 3]). Our simulations were essentially similar to the ones used by Scharef et al. [6]. Our research, however, differed from theirs because it had many more agents and took the distance and communication cost between agents into consideration. These features complicated our experiments but suggested more realistic situations that could express non-simple phenomena compared with previous research. Scharef et al. assumed the environment in which agents resided to be a small-scale local-area network [6], although our experimental setting assumed a large-scale open environment such as the Internet. We think that some features at the critical situation ($tl = 4$) are only observed in a large-scale MAS. [1] discussed a strategy to determine where mobile agents should migrate for efficient and balanced processing. More recently, [2] evaluated very simple search strategies within a peer-to-peer context in given networks to find agents with the required resources. However, these experimental settings differed from ours since our agents do not move but can make autonomous decisions on partner selection based on acquired information. The authors also discussed what kind of structures the network of agents will converge using the same simulation environment in [7], which suggested that learning results were much different depending on workloads. This paper, however, focused on intereference by adaptive selection of collaborating agents in multi-agent systems.

## 5 Conclusion

We investigated how learning parameters for agent local strategies to select partner agents doing a specific task influenced the overall performance of an entire MAS. We used average-value and update functions, which are often used in reinforcement learning, to calculate the statistical values to evaluate all known servers. Although our experiments revealed that the update function could make agents more adaptive, this ability is conducive to constantly changing environments, worsening the overall performance. However, adaptability is vital because the Internet is intrinsically a changing environment. We intend to explore the relationship between the speed of change and adaptation in our next research.

# References

[1] Chow, KP, Kwok, YK (2002) On Load Balancing for Distributed Multiagent Computing. IEEE Transactions on Parallel and Distributed Systems **13**(8) 787–801

[2] Dimakopoulos, V, Pitoura, E. (2003) Performance Analysis of Distributed Search in Open Agent Systems. In: Proceedings of International Parallel and Distributed Processing Symposium (IPDPS'03)

[3] Mehra, T, Wah, BW (1993) Population-based learning of load balancing policies for a distributed computer system. In: Proceedings of Computing in Aerospace 9 Conference. 1120–1130

[4] Mirchandany, R, Stankovic, J (1986) Using stochastic learning automata for job scheduling in distributed processing systems. Journal of Parallel and Distributed Computing **38**(11) 1513–1525

[5] Paxson, V, Floyd, S (1995) Wide-Area Traffic: The Failure of Poisson Modeling. IEEE/ACM Transactions on Networking **3**(3) 226–244

[6] Schaerf, A, Shoham, Y, Tennenholtz, M (1995) Adaptive Load Balancing: A Study in Multi-Agent Learning. Journal of Artificial Intelligence Research **2** 475–500

[7] Sugawara, T, Kurihara, S, Hirotsu, T, Fukuda, K, Sato, S, Akashi, O (2006) Total Performance by Local Agent Selection Strategies in Multi-Agent Systems. In: Proceedings of 5th Int. Joint Conf. on Autonomous Agents and Multiagent Systems (AAMAS2006)

[8] Sugawara, T, Kurihara, S, Hirotsu, T, Fukuda, K, Sato, S, Akashi, O (2006) Adaptive Agent Selection in Large-Scale Multi-Agent Systems. In: Proceedings of the Ninth Pacific Rim International Conference on Artificial Intelligence (PRICAI06)

# Effect of grouping on classroom communities

TORIUMI Fujio and ISHII Kenichiro

Graduate School of Information Science, Nagoya University
Furo-cho, Chikusa-ku, Nagoya City, Aichi Prefecture, Japan
{tori, kishii}@is.nagoya-u.ac.jp

**Summary.** The management of student relationships is one of the most important duties of teachers. Such management usually reflects the teacher's abilities and experiences. The purpose of this study is to clarify management methods to realize appropriate classroom community structures. In this paper, the influence of teacher interventions is analyzed by a classroom community model based on a network model through multi-agent simulations. The proposed model is applied to two simulations. Simulation results suggest the following. Grouping in a classroom community effectively decreases isolated and fringe students. In standard Japanese junior high school classes, the appropriate number of groups is 4 to 6.

**Key words:** Multi-agent simulation, classroom management, social network

## 1 Introduction

In the education field, the management of student relationships is one of the most important duties of teachers. Such management usually reflects the teacher's abilities and experiences. However, it is difficult to clarify the management method by trial and error.

The purpose of this study is to clarify management methods to realize appropriate classroom community structures. In this paper, we propose a classroom community model based on a network model through multi-agent simulations. The proposed method is based Heider's balance theory to analyze the dynamics of group formation.

The multi-agent simulations used in this study are applied to analyze market phenomena [1] [2], societies [3], education fields[4], and so on. Many studies have existentially analyzed the influence of teacher intervention from actual efforts (for example, [5]). On the other hand, few studies have analyzed the influence of teacher intervention by simulation.

In this paper, the influence of teacher interventions is analyzed by multi-agent simulation. We focus on "grouping" and "isolated students." "Grouping" is one method to manage student relationships that is often used in

T. Fujio and I. Kenichiro: *Effect of grouping on classroom communities*, Studies in Computational Intelligence (SCI) **56**, 231–243 (2007)
www.springerlink.com

classroom management [6]. "Isolated students" is one of the most important classroom management issues. The aim of this study is to clarify the most effective method of "grouping" and to suggest a platform on which to erect appropriate class management.

## 2 Proposition of classroom community model

### 2.1 Network model of classroom community

Through communication, people exchange opinions and often alter their attitudes to reflect such obtained information and opinions. Continual communication provokes various types of relationships between people, such as friendship or hostility; through communication people form communities. In this model, a network approach is employed where people are nodes (agents), relationships are links, and a community is a network (Figure 1).

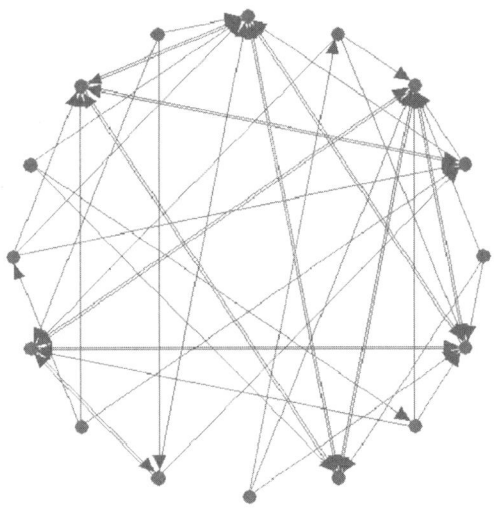

**Fig. 1.** Classroom Community Model

When $m$ people belong to a community, we describe each person as an agent $a_i(i = 1, \cdots m)$. In a network model each agent is described as a node.

We describe a relationship between agents $a_i$ and $a_j(i \neq j)$ as directed link $L_{ij}$. Each link has weight $l_{ij}$ that represents the strength of the feelings of friendship from agent to agent; it is a real number as follows:

$$-1 \leq l_{ij} \leq +1. \tag{1}$$

The greater the value of friendship $l_{ij}$, the higher is the impression of agent $a_i$ for agent $a_j$, and a value of 0 means that agent $a_i$ has no interest in agent $a_j$.

## 2.2 Communication model

In this simulation, friendship and attitudes toward a partner are updated based on Heider's "psychology of interpersonal relations" [7], which is often called "Heider's balance" (HB). This very popular theory describes the dynamics of human relationships.

The attitude of one person toward a subject is affected by the relations among the person himself (P), the subject (X), and communication partner (O). Each of the three opinions, PO, PX, and OX, is expressed as "+" when favorable and "−" when unfavorable. When the sign of the product of these three opinions is positive, the relations preserve the balance. On the other hand, when the sign of the product is negative, the relations are out of balance (Figure 2). According to Heider's theory, when the relations are out of balance, the person tries to "change one's opinion to the subject (PX)" or "change one's remark to the partner (PO)" to achieve balance.

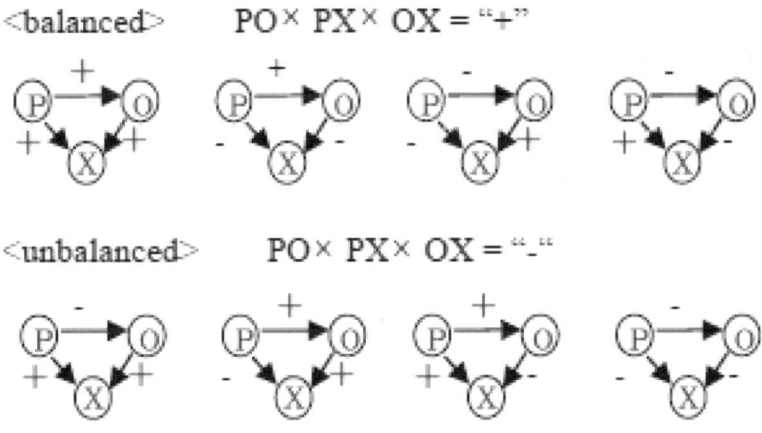

**Fig. 2.** Heider's "psychology of interpersonal relations"

HB usually expresses each PO, PX, and OX opinion as ±1. However, relations between human beings vary in strength, not only in integers [8]. Thus, we adapt a new approach to balance theory that uses real numbers instead of integers [9]. In this approach, the relationship between agents is expressed by real number $R$,

$$-R \le r_{ij} \le R. \tag{2}$$

This is equivalent to eq. (1) when $R = 1$ and $r_{ij} = l_{ij}$.

In this simulation, interpersonal relations are represented as follows.

- PO: Friendship toward partner $l_{ij}$
- PX: Opinion of subject $l_{ik}$
- OX: Partner's opinion of subject $l_{jk}$

When the relation is out of balance, the agent changes both PO and PX to achieve it again. In this simulation, even when relations are balanced, the agent changes PO and PX to strengthen the balance because individuals tend to change their minds when receiving new information from partners. In both balanced and unbalanced cases, the agent changes PO and PX with the following differential equations:

$$\frac{dl_{ij}}{dt} = w \cdot l_{ik} l_{jk} \tag{3}$$

$$\frac{dl_{ik}}{dt} = w \cdot l_{ij} l_{jk}, \tag{4}$$

where $w$ represents the rate of change.

### 2.3 Simulation process

Agents communicate with each other once a unit time (turn) and dynamically change their relationships.

The communication cycle is given below:

1. Communication partner selection
2. Subject selection
3. Communication

### Communication partner selection

Agents select the following agents as communication partners.

- Agents linked by agent $a_i$
- Agents with no partner yet.

Then select agent $a_j$ that has the largest friendship $l_{ij}$ from the candidates. When there are no candidates, agent $a_i$ finishes the current turn without any communication.

## Subject selection

In a new approach to balance theory [9], subject X becomes agent $a_k (k \neq i, j)$. People prefer to talk about interesting topics. This interest is described as the absolute value of friendship $l_{ik}$ because people often talk not only about their friends but also about people they don't really like.

By considering the above assumption, the following equation shows the probability that agent $a_i$ selects agent $a_k$ as a subject of communication:

$$P_k = \frac{|l_{ik}|}{\sum_{p=1}^{m} |l_{ip}|}.$$  (5)

## Communication

Agent $a_i$ changes friendships $l_{ij}$ and $l_{ik}$ by equations (3)(4), when agents $a_i$ and $a_j$ communicate about agent $a_k$. Agent $a_j$ also changes friendships $l_{ji}, l_{jk}$ by equations (3)(4).

After all communications are finished, each agent selects a few agents as friends from the higher order of friendship $l_{ix}$ and then creates new link $L_{ix}$ (Fig. 3). The network is updated from new relations and finishes the current turns.

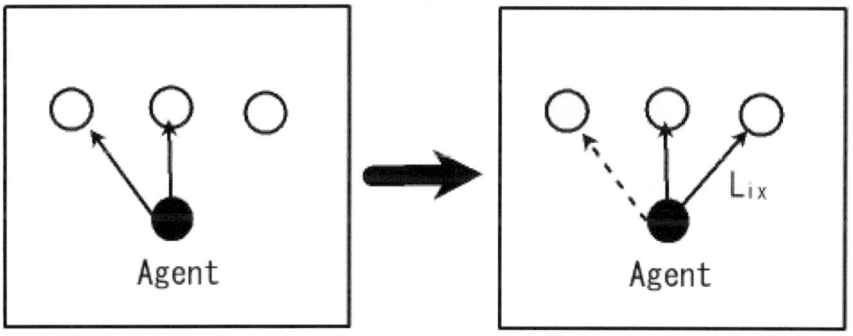

**Fig. 3.** Changing friend links

# 3 Exploratory simulation

## 3.1 Purpose

We clarify adequate simulation turns to analyze the structure of the proposed classroom community model by an exploratory simulation.

Generally, analyzing community structure is difficult at an initial stage due to the drastic changes of member relationships. We analyze a classroom community network after the network state has stabilized.

We adopt "network change rate" as a criterion of network stability. The network change rate is a ratio of changed links in target turns.

The change rate of network $Cr$ is defined by the following equation:

$$Cr_t = \frac{d(t, t-v)}{L_t + L_{t-v}}, \qquad (6)$$

where
$d(t_i, t_j)$ : Number of different links between turns $t_i, t_j$
$L_t$      : Number of links in turn $t$
$v$        : Interval

## 3.2 Simulation conditions

In this simulation, agents are modeled as 30 students. The number of students is based on the average number of students in Japanese junior high schools [10].

The interval for computing the network change rate is 10. We did 100 simulations and used the average.

Exploratory simulation conditions are listed in Table 1.

**Table 1.** Exploratory Simulation Conditions

| | |
|---|---|
| Number of agents | 30 |
| Interval turns $v$ | 10 |
| Number of simulations | 100 |
| Communication period | 10000 |

## 3.3 Simulation results

Figure 4 shows the average "network change rate" at each turn. In the figure, the x-axis shows passing turns, and the y-axis shows average network change rate $Cr$.

From the figure, network change rates peak at 400-600 turns and become smaller than 0.01 after 5000 turns. Therefore, we estimate that network structures become stable at a minimum of 5000 turns after communication starts.

The following simulation communication periods are set to 5000 turns.

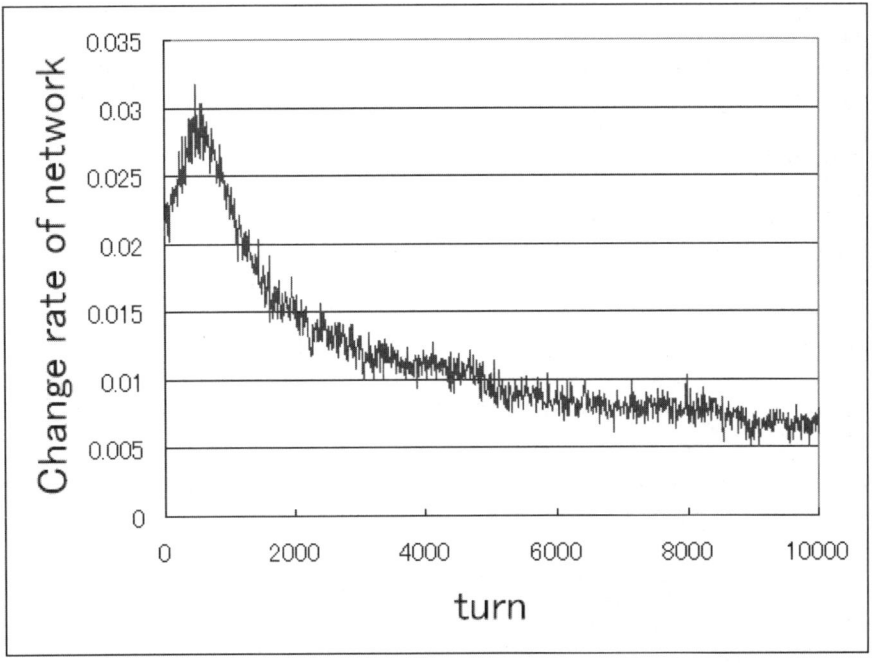

**Fig. 4.** Network change rate

## 4 Effect of grouping in classroom community

### 4.1 Simulation purpose

In experiments, we propose a "grouping" model to analyze the effect of teacher interventions in classroom communities.

In the basic model, agents only select friends as communication partners. However, students are not permitted to only communicate with friendly students. Many chances to communicate with non-friendly students are caused by teacher interventions.

Such communication results in the following effects [6]:

- Growth of student partnerships
- Promotion of friendships
- Realization of new friendships

In this simulation, we focus on the effect of "grouping," which is one frequently used classroom management method. We evaluate the influence of "grouping" on community formation by agents.

## Grouping

To represent grouping, we define "group communication turn." A "group communication turn" occurs in a given probability. When it occurs, all agents perform "group communication" in the turn. "Group communication" is defined as communication between a group members.

Initially, all agents are arranged in groups. At the "group communication turn," agents do not communicate with friendly agents but with agents who belong to the same groups. Hereafter, we call the frequency of the "group communication turn" the "rate of group communication".

## Criteria

The following three parameters are commonly used to evaluate classroom structure:

- Number of isolated agents
- Number of fringe agents
- Cluster coefficient [11]

In this simulation, isolated agents are defined as "agents without input/output links," and fringe agents are defined as "agents without input links" (Fig. 5).

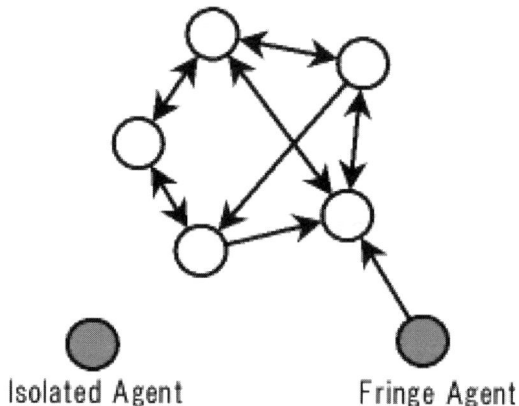

**Fig. 5.** Definitions of isolated and fringe agents

## 4.2 Grouping effect

### Simulation purpose

The purpose of this simulation is to confirm whether grouping effectively decreases the number of isolated and fringe agents.

## Simulation conditions

Simulation conditions are listed in table 2. In this simulation, the rate of group communication increased from 0 to 100% in 5% increments.

**Table 2.** Grouping Simulation Condition

| | |
|---|---|
| Number of agents | 30 |
| Number of group members | 6 |
| Communication periods | 5000 |
| Number of simulations | 100 |

## Simulation results

Figure 6 shows the changes in the numbers of isolated and fringe agents, when the rate of group communication increases. The number of isolated agents decreased drastically when the rate of group communication increased from 0 to 10%. In addition, the number of fringe agents decreased when the rate of group communication increased from 0 to 30%.

Figure 7 shows each network's cluster coefficient, which is increased by group communication.

The results suggest that grouping effectively decreases the number of isolated agents, even if there are few chances for group communication.

### 4.3 Influence of number of groups

## Simulation purpose

Next, we analyze the influence of the number of groups. The purpose of this simulation is to clarify the most appropriate number of groups in a classroom community.

## Simulation conditions

Simulation conditions are listed in Table 3. The group communication rate is set to 50% to predict enough grouping effect.

In this simulation, the number of groups is changed from 2 to 15. The number of group members ranges from 2 to 15.

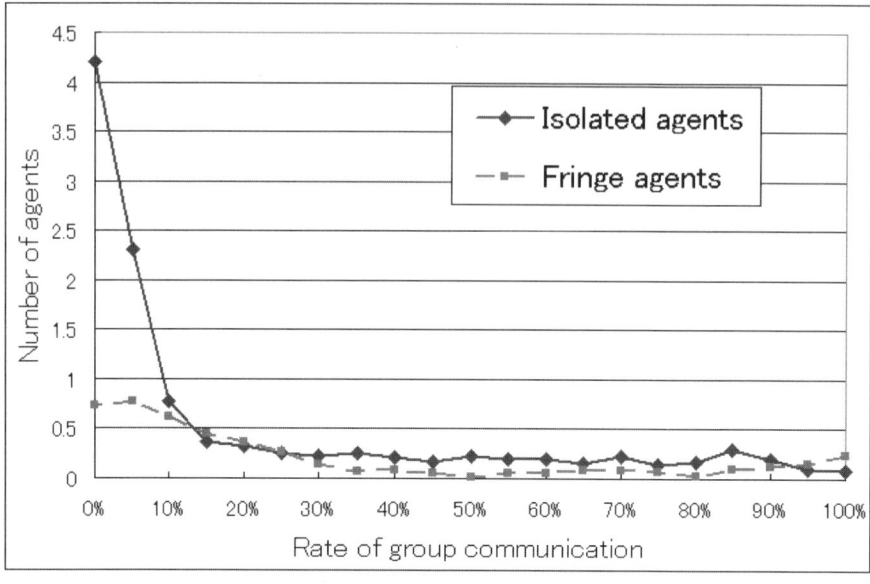

**Fig. 6.** Number of isolated and fringe agents

**Table 3.** Grouping Simulation Conditions

| Number of agents | 30 |
|---|---|
| Group communication rate | 50% |
| Communication periods | 5000 turns |
| Number of simulations | 100 |

### Simulation results

Figure 8 shows the changes of the numbers of isolated and fringe agents, when the number of groups changes. The following results are derived from the figure.

- Too many groups cause an increase of isolated agents
- Too few groups cause an increase of fringe agents

These results suggest that the appropriate number of groups in a standard Japanese junior high classroom is 4 to 6. Thus, the appropriate number of group members is 5 to 8.

In actual classrooms, it often seems that groups have about 5 or 6 members. This simulation ensures the effectiveness of the actual grouping method.

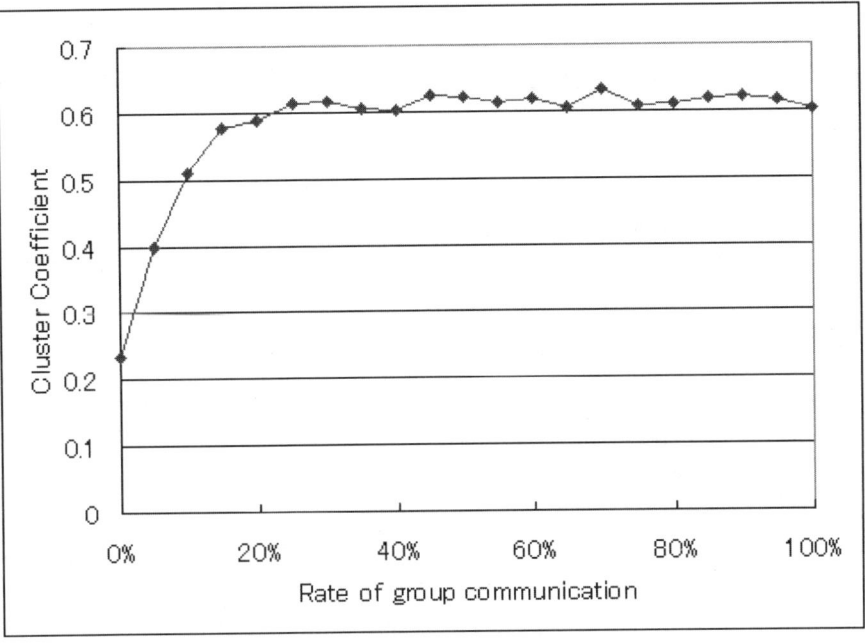

**Fig. 7.** Cluster coefficient

## 5 Conclusion

In this study, we simulated a community-forming mechanism to clarify the influence of grouping on classroom community formation. We represented a community through a communication network using Heider's "psychology of interpersonal relations."

The proposed model is applied to two simulations in which the proposed model is modeled as a standard Japanese junior high school class. Simulation results suggest the following:

- Grouping in a classroom community effectively decreases isolated and fringe students.
- In standard Japanese junior high school classes, the appropriate number of groups is 4 to 6.

The most important future work is confirming the consistency between simulation results and actual classroom problems. We will also try to model other classroom management methods by using the proposed classroom community model, including leadership oriented grouping, individual counseling for isolated students, and so on. Furthermore, considering how to feedback knowledge suggested by the proposed model is another important future work.

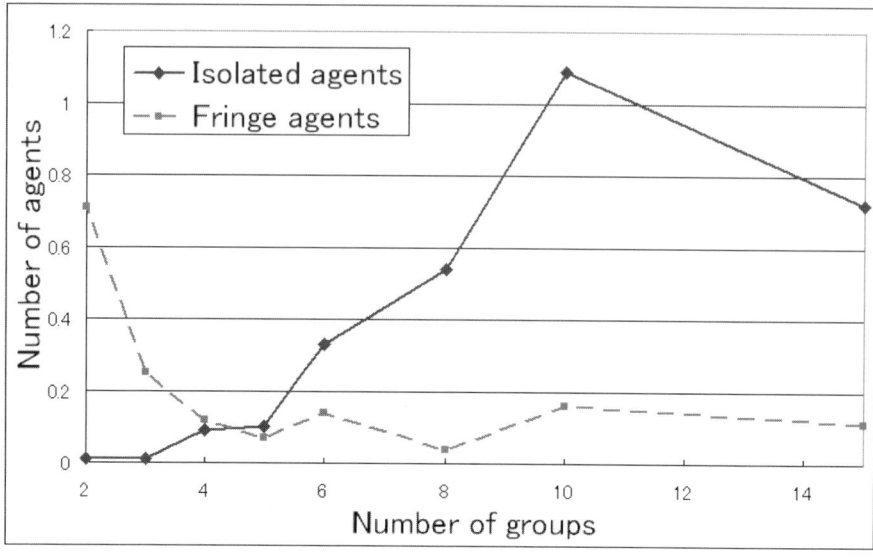

**Fig. 8.** Number of isolated and fringe agents

# References

1. M. Levy, H. Levy, and S. Salomon, Microscopic Simulation of Financial Markets, Academic Press (2000)
2. H. Takahashi, T. Terano, Analysis of Micro-Macro Structure of Financial Markets via Agent-Based Model: Risk Management and Dynamics of Asset Pricing (in Japanese), The Transactions of the Institute of Electronics Information and Communication Engineers of Japan, Vol. J86-D1 No. 8 pp. 618-628 (2003)
3. Joshua M. Epstein and Robert Axtell, Growing Artificial Societies: Social Science from the Bottom Up, Mit Pr (1996)
4. Y. Maeda and H. Imai, An Agent Based Model on the Bully of Mobbed Classmates (in Japanese), The Transactions of the Institute of Electronics Information and Communication Engineers of Japan, Vol. J88-A No. 6 pp. 722-729 (2005)
5. Y. Taruki and T. Ishikuma, Junior High School Students' Group Experiences in Classroom Preparation of Dramas for School Festivals (in Japanese), Japanese Journal of Education Psychology, Vol. 54, pp. 101-111 (2006)
6. The Japanese Society for Life Guidance Studies, Kodomo Shuudan dukuri nyumon (in Japanese). MeijitoshoshuppanKK (2005)
7. F. Heider, The Psychology of Interpersonal Relations, J. Wiley and Sons, New York (1958)
8. Z. Wang and W. Thorngate, Sentiment and social mitosis: Implications of Heider's Balance Theory, J. Artificial Societies and Social Simulation, Vol. 6, No. 3 (2003) (http://jasss.soc.surrey.ac.uk/6/3/2.html)
9. Krzysztof Kulakowski, Przemyslaw Gawronski and Piotr Gronek. "The Heider balance - a continuous approach," Physics, C, 16, 707-716, 2005

10. Ministry of Education, Japan's Education at a Glance 2005 (in Japanese), National Printing Bureau (2005)
11. Watts, D.J. and Strogatz, S.H.: Collective dynamics of 'small-world' networks. Nature 393 440-442 1998

# Emergence and Evolution of Coalitions in Buyer-Seller Networks

Frank E. Walter, Stefano Battiston, and Frank Schweitzer

Chair of Systems Design, ETH Zurich
8032 Zurich, Switzerland
{fewalter, sbattiston, fschweitzer}@ethz.ch

**Summary.** We investigate the dynamics of the creation, development, and breakup of social networks formed by coalitions of agents. As an application, we consider coalition formation in a consumer electronic market. In our model, agents have benefits and costs from establishing a social network by participating in a coalition. Buyers benefit in terms of volume discount and better match of their preferences. Sellers benefit in terms of better predictability of sales volumes.

The model allows us to investigate the stability and size of the coalitions as well as the performance of the market in terms of utility of the agents. We find that the system exhibits three different dominating regimes: *individual purchasing behaviour*, i.e., no social network exists among the agents, *formation of several heterogenous coalitions*, i.e., a number of social networks which are not connected, as well as *condensation to a giant coalition*, i.e., a social network involving all agents.

**Key words:** Coalition Formation, Networks, Multi-Agent Simulation

## 1 Introduction

For several years already, people carry out an increasing amount of their everyday activities on the Internet. For example, people surf the World Wide Web for information, they use Email for communicating, and they buy and sell things on online platforms. One of the core reasons for this increasing ubiquity of the Internet is that it provides the opportunity to form *spontaneous, location-independent communities* in different contexts. In particular, this applies to electronic markets, as can be witnessed through the success of auctioning sites such as Ebay. However, despite the increasing ubiquity of the Internet, the concept of *"buying clubs"* – in which several buyers form a coalition to negotiate with a seller for a special price or special features of a product – has been around for several years, but not really become popular

F. E. Walter et al.: *Emergence and Evolution of Coalitions in Buyer-Seller Networks,* Studies in Computational Intelligence (SCI) **56**, 245–258 (2007)
www.springerlink.com      © Springer-Verlag Berlin Heidelberg 2007

on the Internet. In this paper, we investigate the dynamics of the creation, development, and breakup of social networks formed by coalitions of agents with an example scenario of a consumer electronic market.

We propose that coalition formation may be an alternative means to achieve a trade-off between *economies of scale* and *matching of preferences*. As we can see from the diagram shown in Figure 1, there is a conflict between economies of scale and matching of preferences. Economies of scale occur when mass production of standardised items reduces the costs of producers. Matching of preferences is achieved through the individual production of customised items to match the preferences of consumers. Currently, this conflict leads to the fact that the diversification of products is mostly producer-driven in the sense that producers develop products that are then put on the market and evaluated by consumers in the process of market selection. This feedback from consumers again influences the diversification of products, i.e. it makes producers react by adjusting their supply to the demand of the market. Effectively, this leads to some products to be eliminated from the market and further others to be introduced to the market. We believe that coalition formation among consumers provides another approach to influence the diversification of products: if enough customers join, they can propose a product that would suit their preferences to the market. This would make the process of diversification of products more consumer-driven than it is at the moment. This is desirable both from the perspective of producers – as they get to know the preferences of buyers better – and from the perspective of consumers – as they can let sellers know their preferences and bargain for discounts if many of them have the same preferences.

This paper is structured as follows: following the introduction, in Section 2 we give an overview of the related work done in this area. Subsequently, in Section 3, we describe our model of coalition formation in buyer-seller networks in more detail. Following that, we discuss the results obtained from running multi-agent simulations on this model in Section 4. Finally, we finish the paper with a set of possible extensions in Section 5 as well as a number of concluding remarks in Section 6.

## 2 Related Work

There has been some related work in the field of coalition formation, especially by agents in electronic markets; in the following, we give a brief overview of the related work:

Tsvetovat and Sycara [1] have "reported on coalition formation as a means to formation of groups of customers coming together to procure products at a volume discount." They have provided an incentive analysis for the formation of "buying clubs", looking at the scenario both from the perspective of customers and suppliers. Additionally, Tsvetovat and Sycara have provided

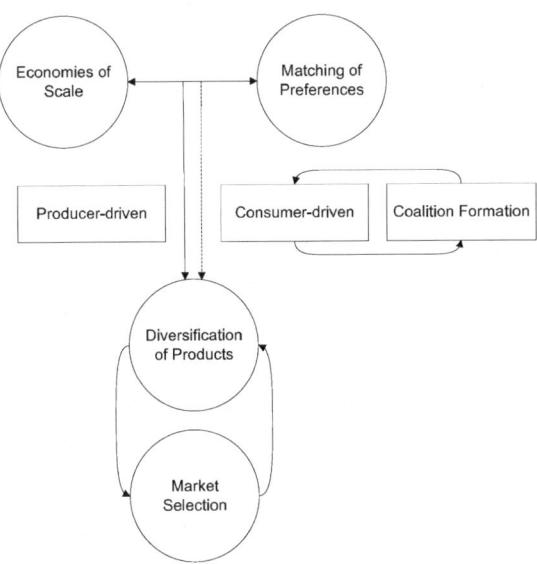

**Fig. 1.** Coalition Formation as an alternative to achieve a trade-off between Economies of Scale and Matching of Preferences

an overview of coalition models, with emphasis on negotiation protocols and cost/utility distribution among agents.

Yamamoto and Sycara [2] have subsequently proposed a coalition formation scheme which they have shown to be stable and efficient.

Furthermore, He and Ioerger [3] have combined the concepts of coalitions and "bundle search". Bundle search occurs when a set of products is to be purchased as a bundle; it is difficult to find the optimal bundle in terms of cost. He and Ioerger propose coalitions as a means of minimising the cost of "bundle search".

Finally, Sarne and Kraus [4] discuss the impact of coalition formation on search cost. They argue that being able to share the cost of searching for a particular product is yet another feature of coalitions. On one hand, the search cost is split among many agents, and on the other hand, the search horizon is widened the more agents are in the coalition.

## 3 Model Description

In the following, we will outline an agent-based model for quantitative analysis of the emergence and evolution of coalitions in buyer-seller networks. Our focus is on the dynamics of creation, evolution, and breakup of coalitions of buyers to obtain a certain product from a seller. We are particularly interested in the effect of *heterogeneity* of preferences of agents on the *size, number*, and *lifetime of coalitions*.

In the model, there are agents which have a set of preferences for the various products they would like to purchase. The goal of the agents is to purchase a product in such a way that there is a high match of its preferences and at the same time a low cost. There is a conflict between these two components, hence agents have to come to a compromise which reflects the best trade-off that they can achieve in this conflict. At each time, each agent has several options: it can engage in an individual purchase, it can initiate a coalition, it can join a coalition, or it can wait for some time and postpone its decision. Each of these options has a unique set of advantages and disadvantages.

One of the core components of the model is the *"more buyers, lower cost" principle*: based on a limited selection of products, buyers have to *compromise* on a product offered by a seller; at the same time, for an increasing number of buyers in a coalition, the cost of the product will decrease. Furthermore, a coalition of buyers can request a customised product from a seller in the hope that this product will be manufactured upon their request. It is important to note that in order for a coalition to succeed, a coalition size threshold has to be reached – only if there are at least this many buyers, a seller will be willing to negotiate with them. Thus, the concept of "buying clubs" has an additional overhead over individual purchasing in form of a waiting time and a risk of not concluding a deal if the coalition size threshold is not reached.

### 3.1 Buyers, Sellers, and Products

In the model, we consider two types of agents, buyers $B$ and sellers $S$. Sellers are represented by their products $P$. Each product $j$ – and thus, also each seller – is characterised by a vector of *features* $[w_{j,1}, ..., w_{j,k}]$ where $w_{j,1...k} \in [0,1]$. Each feature corresponds to a particular property of a product. For example, if the product is a computer, a number of features could be the processor speed, the memory size, the graphics card, and the hard-disk size.

Equally, each buyer $i$ has a set of *preferences* for product features, represented by $[v_{i,1}, ..., v_{i,k}]$ where $v_{i,1...k} \in [0,1]$. Each preference corresponds to a particular value that a given buyer would like the corresponding feature of a product to be. Returning the example in which the product is a computer, the preferences would specify the exact values for the processor speed, the memory size, the graphics card, and the hard-disk size that a particular buyer aims for when he obtains the product from a seller.

The distribution of the features and preferences reflects the heterogeneity of the agent population. At the moment, we assume the features and preferences to be selected randomly from a uniform random distribution on the interval $[0,1]$.

During one run, each buyer has to purchase one product from a seller and his goal is to maximise its own utility. At each time step during the run, each agent has several possible actions that it can do and it needs to select one of these. The run is completed once all buyers are set up to purchase their respective product, either individually or in a coalition.

## 3.2 Utility

The agents in the model are *rational* and *self-interested* and aim to maximise their utility over time in the framework of bounded rationality. The benefit of an agent $i$ from the purchase of a product $j$ depends on:

- the distance between the features $w_j$ of the product $j$ and the preferences $v_i$ of the agent $i$: $\Delta_{ij} = |w_j - v_i|$
- the price of the product $j$, which in turn depends on the quantity sold: $p_j = P/N_j^\beta$ (where $\beta$ measures the price elasticity; in the simulations, we chose $\beta = 0.5$)

An agent's utility thus is a compromise between a cheap price and match of preferences:

$$U_i = \sum_{j=0}^{n} U_{i,j} = \sum_{j=0}^{n} \frac{1}{p_j} \left[ 1 - |w_j - v_i| \right] \tag{1}$$

Note that the utility does not depend on whether the agent is part of a coalition or not. That implies that there is *no direct cost* for an agent to join a coalition, such as a "participation fee". Rather, coalitions come with an *indirect cost* which are the commitment to a coalition as well as the risk of failure in case of unsuccessful coalitions which have $U_i = 0$.

## 3.3 Initialisation

Initially, at time $t = t_{\text{start}} = 0$, each seller offers its product with a given specification. Because the preferences of buyers do not exactly match these specification, each agent has to seek a compromise. This is modelled by a parameter $\epsilon$, which measures the flexibility of agents to accept products different from their preferences. More specifically, buyers agree on a certain product if the difference between their preferences and the product features is less than $\epsilon$. If, initially, several agents agree on a certain product, this is seen as a coalition formation. This procedure of mapping agents to coalitions based on their preferences and the product features of the coalitions is illustrated in Figure 2.

## 3.4 Possible Actions

At each further step in time, each agent that has not yet committed to either a coalition or an individual purchase has four different possible actions that

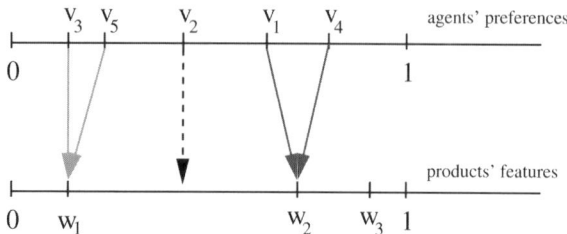

**Fig. 2.** Mapping of Agents to Coalitions by mapping Preferences to Features: agents within a certain distance to a particular product, $\epsilon$, are mapped to a coalition on this product. For illustration purposes, this diagram considers the case of preferences/features with only one dimension

it can take, each with a unique set of advantages and disadvantages as well as a particular rate at which the respective action is performed. The actions are the following:

1. *Purchase product j individually.* This action has the advantage that the agent immediately gets the product with no further waiting time; however, it also has the disadvantage that the agent has to pay a higher price $p_i = P$.

   Thus, the rate of performing this action is

$$k_i^{\text{ind}}(t) \propto \frac{1}{P}\left[1 - \Delta_{ij}\right] \qquad (2)$$

   Hence, the rate of performing this action is higher for lower prices as well as lower distances of the agent's preferences and the product's features.

2. *Join existing coalition j with a set of other buyers $N_j$.* This action has the advantage that the agent pays a lower price, the more agents already have joined the existing coalition, $p_i = P/\sqrt{N_j}$; however, it also has the disadvantages of having to wait until the coalition has reached critical size $N_j \geq N_{\text{thr}}$ with which the transaction will take place as well as the risk of coalition failure if this threshold is not reached.

   Consequently, the rate of performing this action is

$$k_i^{\text{coal}}(t) \propto \frac{\sqrt{N_j}}{P} \frac{N_j}{N_{\text{thr}}}\left[1 - \Delta_{ij}\right] \qquad (3)$$

   Hence, the rate of performing this action increases with the number of agents already in the particular coalition (because this decreases the price as well as increases the likelihood of reaching the coalition size threshold) as well as with better match of preferences (because this decreases the distance between the agent's preferences and the product's features).

3. *Initiate new coalition k and wait for other buyers to join.* This has the advantage that the agent can propose a product $k$ according to its own preferences and thus achieve a perfect match of its preferences $\Delta_{ik} = 0$; however, there still are the disadvantages of risking a coalition failure as $N_k(t_0) = 1 \ll N_{\mathrm{thr}}$ as well as having to wait until the coalition has reached the critical size.

Thus, the rate of performing this action is

$$k_i^{\mathrm{init}}(t) \propto \frac{\sqrt{1}}{P} \frac{1}{N_{\mathrm{thr}}} \qquad (4)$$

Hence, the rate of performing this action is lower for higher coalition size thresholds and higher prices.

4. *Postpone decision.* The advantage of this wait-and-see action is that the agent does not commit itself and is open for future possibilities, i.e. in the form of additional coalitions that match its preferences well. However, there are also the disadvantages of having to wait as well as the future being uncertain, i.e. it is by no means guaranteed that future coalitions will match its preferences well.

The rate of performing this action is

$$k_i^{\mathrm{wait}}(t) \propto \exp(-\alpha t) \qquad (5)$$

i.e. it decreases exponentially with time (where $\alpha$ is a parameter controlling the exponential decrease).

### 3.5 Temporal Structure and Decision Dynamics

We assume a *discrete linear bounded model of time.* During each run of the simulation, each buyer has to purchase one product from a seller – either individually or in a coalition. Each run is divided into time steps, at each time step $t$, each agent has to make a decision as to whether it would like to purchase its product individually, by joining a coalition, by initiating a coalition, or by waiting and postponing the decision to a later time.

Thus, at each time step $t$, the agent computes the rates of performing a particular action for each of the possibilities $k_i^{\mathrm{ind}}(t)$, $k_i^{\mathrm{coal}}(t)$, $k_i^{\mathrm{init}}(t)$, and $k_i^{\mathrm{wait}}(t)$ for each of the actions (note that there may be more than one action of each type, e.g. several individual purchase possibilities or several existing coalitions that can be joined). Thus, each possible action has a certain weight $k_i$. To make a decision for one of these actions, the agent performs a stochastic draw among the weighted possibilities. This is illustrated in Figure 3.

A run is completed once no more buyers are waiting and each one has selected a way of purchasing its respective product, either individually or in a coalition.

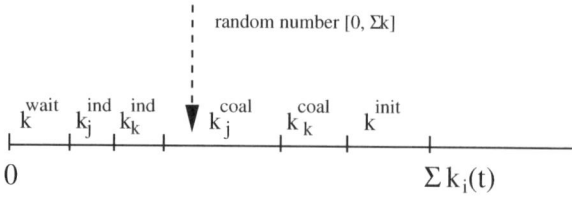

**Fig. 3.** Decision Dynamics: stochastic draw among all possible actions

The consequences of this temporal structure and decision dynamics are that there is a symmetry break through the stochastic draw among several possible actions. This is coupled with positive feedback as the the decision of one agent $i$ affects the weights $k_j$ of other agents.

In repeated games – which we do, in the current model, not yet consider – the utility of an agent at deal completion time $t = t_{\text{end}}$ in one run should affect the strategy of that agent in the next run: i.e. an agent is more likely to deal with other agents that he has made a good experience with. Such behaviour will further investigated in extensions of the model discussed in this paper.

## 4 Results

In this section, we discuss the results obtained through computer simulations of the model. The core observation made is that there are three dominating regimes, namely individual purchasing behaviour, formation of several hetero-

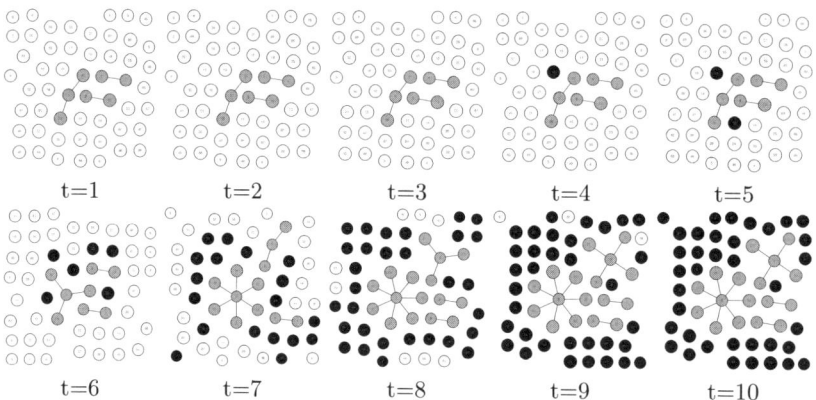

**Fig. 4.** Individual Purchasing Behaviour: evolution of the network of agents in one run over time, from $t = t_{\text{start}} = 0$ before any decisions to $t = t_{\text{end}} = 10$ when all agents have decided for either initiating/joining a coalition or engaging in an individual purchase. In the example, $N_{\text{sellers}} = N_{\text{products}} = 3$, $\epsilon = 0.01$ and $N_{\text{thr}} = 50$

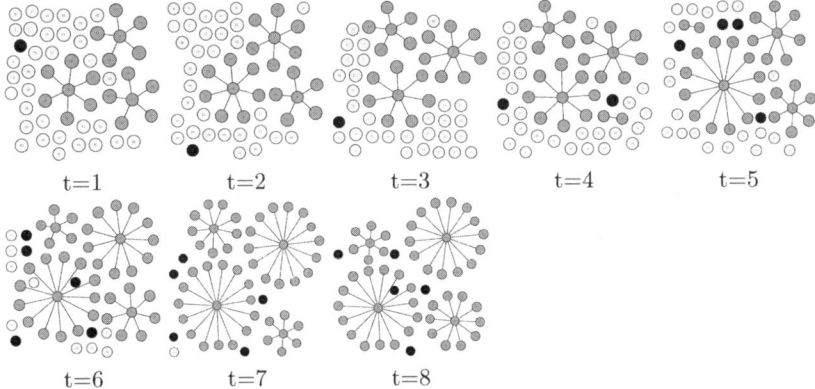

**Fig. 5.** Formation of Several Heterogeneous Coalitions: evolution of the network of agents in one run over time, from $t = t_{\text{start}} = 0$ to $t = t_{\text{end}} = 8$ when all agents have decided for either initiating/joining a coalition or engaging in an individual purchase. In the example, $N_{\text{sellers}} = N_{\text{products}} = 3$, $\epsilon = 0.05$ and $N_{\text{thr}} = 5$

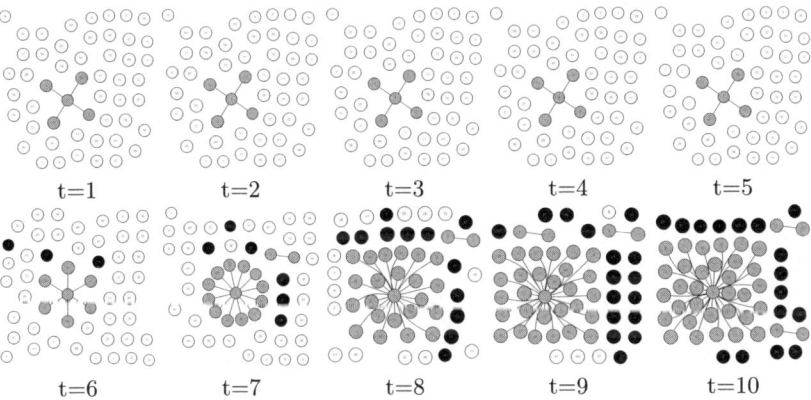

**Fig. 6.** Condensation into a Giant Coalition: evolution of the network of agents in one run over time, from $t = t_{\text{start}} = 0$ to $t = t_{\text{end}} = 10$ when all agents have decided for either initiating/joining a coalition or engaging in an individual purchase. In the example, $N_{\text{sellers}} = N_{\text{products}} = 1$, $\epsilon = 0.05$ and $N_{\text{thr}} = 25$

geneous coalitions, and condensation into a giant coalition. We discuss the observations for each of these regimes in the following subsections.

In our simulations, we found these results for varying agent populations. Usually, we ran simulations for large agent populations (e.g. simulations with 500 agents, averaged over at least 500 runs) for statistically significant results, but we have, for illustrative purposes, included plots of the evolution of the network for smaller – and thus visualisable – agent populations (e.g. 1 run of a simulation with 50 agents each). These are shown in Figures 4, 5, and 6, for the three dominating regimes, respectively. In these graphs, white nodes correspond to yet undecided or waiting agents, black nodes to agents who have committed to an individual purchase, and gray nodes to agents who have committed to a coalition. Thus, initially, at $t = t_{\text{start}}$, the whole agent population will be white and at $t = t_{\text{end}}$, the whole agent population will be either gray or black. A gray node with a blue label (instead of the red label of agents) indicates the product a coalition is on.

The three regimes arise for different assignments of values for the variables $N_{\text{sellers}} = N_{\text{products}}$, $\epsilon$, and $N_{\text{thr}}$.

### 4.1 Individual Purchasing Behaviour

In the case of *individual purchasing behaviour*, most agents do not form coalitions, but rather buy individually. Of course, it may be that there is a minority of agents that either initiate or join coalitions, but the majority of agents engages in individual purchases. Virtually none of the coalitions reach the threshold for a successful coalition.

Figure 4 shows the evolution of the network of agents in one run over time. From the sequence of snapshots it can be seen that initially, agents are reluctant to come to a decision, but after a few steps have passed, more and more agents decide for an individual purchase.

### 4.2 Formation of Several Heterogeneous Coalitions

In the case of the *formation of several heterogeneous coalitions*, agents form several coalitions of various sizes. There is a minority of agents that engages in individual purchases, but the majority of agents either initiates or joins coalitions. A vast fraction of these coalitions succeeds.

Figure 5 shows the evolution of the network of agents in one run over time. From the sequence of snapshots it can be seen that, over time, agents initiate a number coalitions which then are joined by more and more agents.

### 4.3 Condensation into a Giant Coalition

In the case of *condensation into a giant coalition*, one agent initiates a coalition which is subsequently joined by almost all other agents. There is a minority of agents that either initiates or joins other coalitions or engages in an individual purchase, but the majority of agents takes part in the giant coalition. The giant coalition succeeds.

Figure 6 shows the evolution of the network of agents in one run over time. From the sequence of snapshots it can be seen that initially, there is one coalition which is then joined by more and more agents – the coalition size threshold is so high that the number of coalition initiations is low. Nonetheless, there is a fraction of the agents which decides for individual purchases.

## 4.4 Performance of the System

Apart from the network plots, we have also recorded the corresponding fractions of agents in coalitions and in individual purchases over time, respectively, see Figure 7. From these graphs, it is visible that

- for individual purchasing (a), the majority of agents waits for a few time steps and then makes the decision to engage in an individual purchase; a minority of agents initiates or joins coalitions.
- for several heterogeneous coalitions (b), the majority of agents joins a coalition at about $t = (t_{end} - t_{start})/2$; nonetheless, there is a minority of other agents that engages in individual purchases.
- for a giant coalition (c), the majority of agents waits for some time and then joins the giant coalition; a minority of agents engages in individual purchases. The actual proportion of agents in coalitions versus agents in individual purchases is lower than for the scenario with several heterogeneous coalitions (and obviously significantly higher than for the scenario with individual purchasing).

It should be noted that in each of these regimes, there is a dominating structure, but that in each case, there is a minority of agents that do not follow the majority. Furthermore, the average utility of an agent in the system is highest for the scenario in which there are several heterogeneous coalitions of various sizes. An interpretation for this behaviour is that in this regime, the success rate of coalitions is quite high (because they usually have a size greater than the threshold) and, at the same time, the diversity of the different coalitions leads to quite good preference matching (because each agent finds a coalition it can join within an acceptable distance).

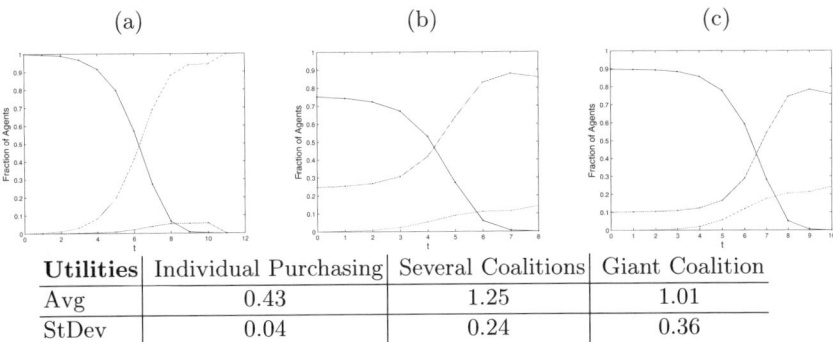

| Utilities | Individual Purchasing | Several Coalitions | Giant Coalition |
|---|---|---|---|
| Avg | 0.43 | 1.25 | 1.01 |
| StDev | 0.04 | 0.24 | 0.36 |

**Fig. 7.** Fraction of Agents in Coalitions vs. Time (black lines: waiting agents; red lines: agents in individual purchases; blue lines: agents in coalitions)

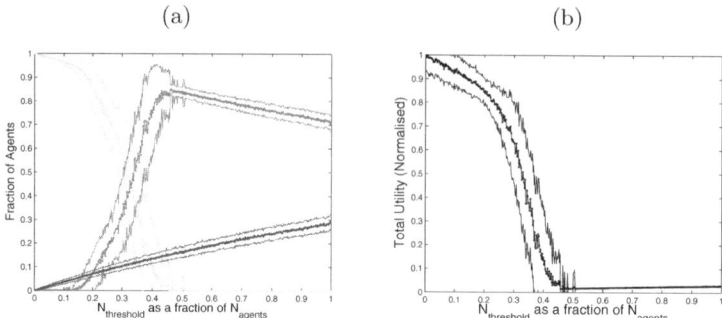

**Fig. 8.** (a) the fraction of agents in successful coalitions (green), in unsuccessful coalitions (red), and in individual purchases (blue) against $N_{\mathrm{thr}}$ as a fraction of $N_{\mathrm{agents}}$ and (b) the total utility of agents normalised to a value in the interval $[0, 1]$, also against $N_{\mathrm{thr}}$ as a fraction of $N_{\mathrm{agents}}$. All plots show the standard deviation from the mean over 500 runs

Figure 8 shows two further plots of (a) the fraction of agents in successful coalitions, in unsuccessful coalitions, and in individual purchases against $N_{\mathrm{thr}}$ as a fraction of $N_{\mathrm{agents}}$ and (b) the total utility of agents normalised to a value in the interval $[0, 1]$, also against $N_{\mathrm{thr}}$ as a fraction of $N_{\mathrm{agents}}$. From these graphs, it is clearly visible that for increasing $N_{\mathrm{thr}}$ as a fraction of $N_{\mathrm{agents}}$, the number of successful coalitions decreases. At the same time, for increasing $N_{\mathrm{thr}}$ as a fraction of $N_{\mathrm{agents}}$, the number of individual purchases increases as well. Furthermore, the total utility of agents decreases for increasing $N_{\mathrm{thr}}$ as a fraction of $N_{\mathrm{agents}}$.

## 5 Possible Extensions

In this paper, we have described the basic version of a coalition formation model. It is possible to extend this model to investigate various other aspects of coalition formation. Some of these possible extensions are repeated games, i.e. simulating several runs of the same set of agents in sequence, as well as modifications to the buyers' and sellers' dynamics. We plan to implement all these changes in further versions of the model.

In repeated games, it would be possible to investigate the behaviour of buyer and seller agents over a number of iterated runs. Buyers would have a memory to store information about the failure or success of coalitions; equally, sellers would have a memory to keep track which buyers they dealt with (to reward loyal behaviour, for example) or which other sellers offer similar products and at which conditions (to be able to create special offers to bait for agents). Thus, the memory allows agents to incorporate feedback from their past experiences into future decisions. Thus, over time, trust relationships between agents can develop [5]. In this framework, it would be interesting whether stationary or non-stationary coalitions form and sustain themselves over time.

Similarly, it would be possible to modify the buyers' and sellers' dynamics in various ways: for keeping the scenario simple, we currently consider the preferences and features to be one-dimensional; in a more complex scenario, these could be multi-dimensional, with individual components of the preferences and features having different weights (i.e. importance) for different agents, and with the values evolving over time. Further, if buyers only know incomplete, bounded in time information about products as well as if sellers offer more than just one product as in the current model, the scenario would become more realistic. A limited budget for buyers and limited production resources for sellers would be a means to enforce competition.

# 6 Conclusion

In this paper, we have investigated the dynamics of coalitions in the scenario of a consumer electronic market. Coalitions are social networks of agents to reach certain goal – customised products at lower prices. When joining coalitions, agents have to *compromise* between better matching of their preferences and lower prices; they can achieve a higher utility through coalitions, but also face the risk of the failure of a coalition if not enough other agents join it.

We have presented a modelling framework for comparing the formation of coalitions to individual buying behaviour. Our focus was on the impact of the heterogeneity of the preferences of agents as well as of the features of products and on the impact of the coalition size threshold on the overall utility experienced in the system. We have identified three different scenarios – individual purchasing behaviour, formation of several heterogeneous coalitions, and condensation to a giant coalition – and observed that the utility of the system is maximised for the scenario in which several heterogeneous coalitions form. For further versions of the model, we suggest to incorporate extensions towards multiple products as well as multiple features of/preferences for these products, to observe learning effects in repeated games, and to enforce competition scenarios by limiting the resources of both buyers and sellers.

We believe that coalition formation provides a means for *a more consumer-driven approach to a trade-off between economies of scale and matching of preferences*. Coalitions allow for better matching of preferences, more accurate predictions of sales volumes, and reduced costs of interaction between buyers and sellers. All these features make coalition formation attractive for consumers to take more impact on the – currently very producer-driven – process of product diversification.

# References

1. Tsvetovat, M., Sycara, K., Chen, Y., Ying, J.: Customer coalitions in the electronic marketplace. In: Proceedings of the 7th Conference on Artificial Intelligence (AAAI-00) and of the 12th Conference on Innovative Applications of Artificial Intelligence (IAAI-00), AAAI Press (2000) 1133–1134
2. Yamamoto, J., Sycara, K.: A stable and efficient buyer coalition formation scheme for e-marketplaces. In Müller, J.P., Andre, E., Sen, S., Frasson, C., eds.: Proceedings of the Fifth International Conference on Autonomous Agents, ACM Press (2001) 576–583
3. He, L., Ioerger, T.R.: Combining bundle search with buyer coalition formation in electronic markets: A distributed approach through negotiation. In: Proceedings of the Third International Joint Conference on Autonomous Agents and Multiagent Systems (AAMAS 2004), IEEE Computer Society (2004) 1440–1441
4. Sarne, D., Kraus, S.: Buyer's coalition for optimal search. In: Proceedings of the Fourth International Joint Conference on Autonomous Agents and Multiagent Systems (AAMAS 2005), ACM Press (2005) 1225–1226
5. Battiston, S., Walter, F.E., Schweitzer, F.: Impact of trust on the performance of a recommendation system in a social network. In: Proceedings of the Workshop on TRUST at the 5th International Joint Conference on Autonomous Agents and Multiagent Systems (AAMAS-06), (in press) (2006) 1–8

Printing: Krips bv, Meppel
Binding: Stürtz, Würzburg